"十四五"职业教育国家规划教材

高职高专名校名师精品
"十三五"规划教材

Java Web 动态网站开发 微课版

Development of Dynamic Web Site Based on Java Web

张桓 李金靖 ◎ 主编
丁明浩 张昊楠 ◎ 副主编

人民邮电出版社
北京

图书在版编目（CIP）数据

Java Web动态网站开发 : 微课版 / 张桓, 李金靖主编. -- 北京 : 人民邮电出版社, 2019.2(2024.7重印)
高职高专名校名师精品“十三五”规划教材
ISBN 978-7-115-50193-6

Ⅰ. ①J… Ⅱ. ①张… ②李… Ⅲ. ①JAVA语言－网页制作工具－高等职业教育－教材 Ⅳ. ①TP312.8 ②TP393.092.2

中国版本图书馆CIP数据核字(2018)第279972号

内 容 提 要

本书从初学者的角度出发，通过通俗易懂的语言、丰富多彩的案例，详细介绍了进行 Java Web 应用开发应掌握的各方面技术。全书共分 8 个项目，主要内容包括开发环境的搭建，JSP 语法基础，JSP 内置对象的使用，JavaBean 技术的应用，JDBC 数据库接口的使用，Servlet 技术的应用，以及两个综合案例的开发讲解。

本书采用项目驱动方式，所有知识点都结合具体实例进行讲解，尤其是最后的两个综合实训案例都来自作者实际开发的项目。全书内容丰富，系统性和应用性强，融入了作者多年教学和实践的经验及体会。通过学习本书，读者可以轻松了解 Java Web 应用开发的精髓，快速掌握 JSP 的开发技能。

本书既可作为高职高专、高等教育院校计算机及相关专业的教材，也可作为广大 Web 应用开发者自学的入门教材，还可作为从事 Web 应用开发的工程技术人员学习和应用的参考书。

◆ 主　　编　张　桓　李金靖
　副 主 编　丁明浩　张昊楠
　责任编辑　刘　佳
　责任印制　马振武

◆ 人民邮电出版社出版发行　　北京市丰台区成寿寺路 11 号
　邮编　100164　　电子邮件　315@ptpress.com.cn
　网址　http://www.ptpress.com.cn
　三河市君旺印务有限公司印刷

◆ 开本：787×1092　1/16
　印张：15　　　　　　2019 年 2 月第 1 版
　字数：392 千字　　　2024 年 7 月河北第18次印刷

定价：48.00 元

读者服务热线：(010)81055256　印装质量热线：(010)81055316
反盗版热线：(010)81055315
广告经营许可证：京东市监广登字 20170147 号

前言
Foreword

党的二十大报告中指出教育是国之大计、党之大计。培养什么人、怎样培养人、为谁培养人是教育的根本问题。育人的根本在于立德。本书在内容上，采取恰当方式，自然融入中华传统文化、科学精神、职业素养和爱国情怀等元素。注重挖掘 Java Web 学习与工作生活之间的紧密联系，将“为学”和“为人”有机地结合在一起。

经过多年的发展，Java Web 技术已经成为 Web 应用开发的主流技术之一，为越来越多的 Web 应用开发人员所使用。本书以培养读者掌握 Java Web 开发技术的基本能力为主旨，结合作者长期从事 Java Web 教学与开发的实践经验，以精心的项目章节安排与知识体系设计、先进的教学理念，循序渐进地展开教学内容。本书能够使初学者牢固建立起 Web 应用开发的编程理念，为读者进一步学习后续知识体系打下坚实的基础。对于有一定 Java Web 开发基础的读者，本书能够更好地帮助他们梳理知识体系，将各个分散的知识点凝聚到实际 Java Web 项目开发这条主线上来。

本书精心设计了八个思政小故事，因势利导，依据专业课程的特点，采取了恰当方式，自然融入中华传统文化、科学精神、职业素养和爱国情怀等元素。注重挖掘课程中的思政教育要素，弘扬精益求精的专业精神、职业精神和工匠精神。有意识地培养学生的创新精神，将“为学”和“为人”相结合。

全书将 Java Web 应用开发的精髓知识点分解为 8 个项目，划分成两个部分：项目一至项目六为第一部分，围绕 Java Web 应用开发的基础知识点展开，内容包括 Java Web 开发环境的搭建、开发工具的安装与使用、JSP 的基本语法、JSP 内置对象的使用、JDBC 数据库接口技术的应用、JavaBean 技术的应用、Servlet 技术的应用等；项目七至项目八为第二部分，围绕实际 Java Web 项目开发的实现展开，内容包括 JSP + JavaBean 开发模型的“天码行空”企业网站项目的实现、JSP + JavaBean + Servlet 开发模型的孕婴网站项目的实现等。

本书以高职高专计算机相关专业和其他有 Web 应用开发需求的工科专业的初学者为主要对象，也可作为 Java Web 开发人员的参考书。使用本书建议采用理论实践一体化教学模式，参考学时见下面的学时分配表。

学时分配表

项目	课程内容	学时
项目一	Java Web 概述	4
项目二	JSP 基本语法	8
项目三	JSP 内置对象	8
项目四	JDBC 技术的应用	8
项目五	JavaBean 技术的应用	8

续表

项目	课程内容	学时
项目六	Servlet 技术的应用	8
项目七	“天码行空”企业网站的设计与实现	10
项目八	孕婴网站的设计与实现	10
课时总计		64

本书配备了教学项目案例、微课视频、教学课件等学习资源，方便读者在课堂之外继续学习。本书编写中力求重点突出、难易适中，在强调知识原理的基础上，注重思维训练，提高读者的项目开发能力。本书的成稿得益于一支工学结合的编写团队。参与各项目编写的人员均是国家级示范高职院校的一线骨干教师，他们具备丰富的教学经验及项目开发实践经验，了解如何将理论知识转化为实际开发能力。本书由张桓、李金靖任主编，由丁明浩、张昊楠任副主编。其中的项目一和项目七由张桓编写，项目五和项目八由李金靖编写，项目四和项目六由丁明浩编写，项目二和项目三由张昊楠编写，全书由张桓统稿并审核。

在本书的成稿与出版过程中，出版社的编辑同志以高度负责的敬业精神，付出了大量的心血。还有很多同行及专家提出了许多的宝贵意见。在此，对所有提供过帮助的同志表示衷心的感谢！由于作者水平所限，书中难免有不妥之处，敬请各位读者与专家批评指正。

编者

2022 年 11 月

目录
Contents

项目四　JDBC 技术的应用 60

项目五　JavaBean 技术的应用 89

项目六 Servlet 技术的应用 113

项目七 “天码行空”企业网站的设计与实现 138

项目八 孕婴网站的设计与实现 185

PART01

项目一 Java Web概述

JSP是一种Web开发技术，以Java语言为基础，与HTML语言紧密结合，可以实现Web页面设计。目前，JSP技术已成为Web应用开发的主流技术之一，并已广泛应用于电子商务、电子政务、网络资源管理等领域。

课堂学习目标

- 了解JSP开发技术
- 掌握JSP开发环境的搭建
- 掌握JSP页面的基本实现

素养拓展

- 千里之行，始于足下

素养拓展

任务一　Web 开发技术概述

任务要求

本任务要求了解 Web 开发技术的基本知识，认识静态网页与动态网页，了解 Web 服务器等概念。

任务实现

（一）Web 基础知识

WWW 是 World Wide Web 的缩写，也可以简称为 Web，中文名字为“万维网”。Web 出现于 1989 年 3 月，是由欧洲量子物理实验室（CERN）开发出来的主从结构分布式超媒体系统。1990 年 11 月，第一个 Web 服务器正式运行，通过 Web 浏览器可以看到 Web 页面。目前，与 Web 相关的各种技术标准都由万维网联盟（W3C）负责管理和维护。

Web 是一个分布式的超媒体信息系统，它将大量的信息分布在网上，为用户提供更多的多媒体网络信息服务。借助万维网，人们只要通过简单的方法就可以很迅速方便地取得丰富的信息资料。用户通过 Web 浏览器访问信息资源的过程中，无须关心一些技术性的细节，而且界面友好，因而 Web 推出后就受到了热烈的欢迎，并得到了飞速的发展。

从技术层面上看，Web 技术可以分为客户端技术和服务器端技术。

（二）Web 客户端技术

Web 是一种典型的分布式应用架构。Web 应用中的每一次信息交换都要涉及客户端和服务器端两个层面。本节主要介绍 Web 客户端技术。

Web 客户端的主要任务是展现信息内容。Web 客户端技术主要包括 HTML 语言、客户端脚本语言、CSS 样式表及一些衍生技术。

1. HTML 语言

HTML 的全称是 Hypertext Markup Language，即超文本标记语言。它是客户端技术的基础，主要用于显示网页信息，由浏览器解释执行，不需要编译。HTML 简单易用，使用 HTML 可以实现在网页中定义标题、文本、表格或者图片等信息。严格地说，HTML 并不能算作是一种程序设计语言，它缺少程序设计语言所应有的特征。

2. CSS 样式表

CSS 的全称是 Cascading Style Sheet，即层叠样式表。在制作网页时采用 CSS 样式表，可以有效地对页面的布局、字体、颜色、背景等效果实现更加精确的控制。HTML 与 CSS 是“内容”与“形式”的关系，由 HTML 来确定网页的内容，由 CSS 来实现页面的表现形式。CSS 大大提高了开发者对信息展现格式的控制能力。

3. 客户端脚本语言

客户端脚本技术是指嵌入到 Web 页面中的程序代码，这些程序代码是一种解释性的语言，浏览器可以对客户端脚本进行解释。通过脚本语言可以实现以编程的方式对页面元素进行控制，从而增加页面的灵活性。常用的客户端脚本语言有 JavaScript 和 VBScript。目前，应用较为广泛的客户端脚本语言是 JavaScript。

（三）Web 服务器端技术

最早的 Web 服务器只是简单地响应浏览器发来的 HTTP 请求，并将存储在服务器上的 HTML 文件返回给浏览器。现在的服务器端应用技术主要用于进行业务逻辑处理和与数据库等服务进行交互操作等。

Web 服务器端技术主要包括 ASP 技术、JSP 技术、PHP 技术及一些衍生技术。

1. ASP 技术

ASP（Active Server Page）是一种由微软公司提供的，使用很广泛的动态网站开发技术。它通过在页面代码中嵌入脚本语言来生成动态的内容，在服务器端必须安装适当的解释器之后，才可以通过调用此解释器来执行脚本程序。ASP 技术主要用于 Windows 平台中。随着微软 Windows 平台进入.NET 技术体系时代，ASP 技术也发展为 ASP.NET。ASP.NET 是.NET 框架的一部分，可以使用任何.NET 兼容的语言来编写 ASP.NET 应用程序。

2. JSP 技术

JSP（Java Server Page）是以 Java 为基础开发的，所以它沿用了 Java 强大的 API 功能。JSP 页面中的 HTML 代码用来显示静态内容部分，嵌入到页面中的 Java 代码与 JSP 标记用来生成动态内容部分。JSP 可以被预编译，提高了程序的运行速度。JSP 开发的应用程序经过一次编译后，可以运行在绝大部分系统平台中，代码无须做修改。

3. PHP 技术

PHP（Personal Home Page）是一种开发动态网页的技术，是一种开源的 Web 服务器脚本语言。PHP 的语法类似于 C 语言，并且混合了 Perl、C++和 Java 的一些特性。在 PHP 中提供了许多已经定义好的函数，扩展性强。PHP 可以被多个平台支持，但广泛应用于 UNIX/Linux 平台。

任务二　JSP 开发技术概述

任务要求

本任务要求了解 JSP 开发技术的基本知识，认识 JSP 页面的基本组成。

任务实现

（一）JSP 基础

JSP 技术是由 Sun 公司倡导，多家公司参与建立的一种动态网页技术标准。它是在传统的网页 HTML 文件中插入 Java 程序段和 JSP 标记，形成 JSP 文件，后缀名为“.jsp”。

JSP 将网页逻辑与网页设计的显示分离，支持可重用的基于组件的设计，使 Web 应用程序的开发变得迅速和容易。用 JSP 开发的 Web 应用是跨平台的，可以在绝大多数服务器上运行。自 JSP 推出后，众多大公司都推出支持 JSP 技术的服务器，如 IBM、Oracle、Bea 等公司，所以 JSP 迅速成为商业应用的服务器语言。

Web 服务器在遇到访问 JSP 网页的请求时，首先执行其中的程序段，然后将执行结果连同 JSP 文件中的 HTML 代码一起返回给客户。JSP 与 Java Servlet 技术一样，是在服务器端执行的，通常返回客户端的就是一个 HTML 文本，因此客户端只要在浏览器中就能浏览 JSP 页面。

JSP 的主要特点：

➢ 一次编写，到处运行。
➢ JSP 系统获得多平台的支持。
➢ 具有强大的可伸缩性。
➢ 获得多样化和功能强大的开发工具的支持。

（二）JSP 页面的组成

JSP 页面主要包括两个部分：一个是静态部分，如 HTML 标记、CSS 样式等，用来完成信息的显示和样式的控制；另一个是动态部分，如 JSP 指令、嵌入的 Java 代码等，用来完成数据的获取和处理。

JSP 页面元素的组成：

➢ 静态部分：HTML 标记、CSS 样式和普通的静态文本。
➢ 指令：以“<%@ 指令名”标记开始，以“%>”标记结束。
➢ 表达式：<%= Java 表达式%>。
➢ 脚本：<% Java 代码%>。
➢ 声明：<%! 方法或者变量%>。
➢ 动作：以“<jsp：动作名>”标记开始，以“</jsp：动作名>”标记结束。
➢ 注释：<%-- 注释内容 --%>。

任务三　JSP 开发环境的搭建

任务要求

本任务要求了解 JSP 开发环境的组成，掌握 JSP 开发环境的搭建。

任务实现

（一）安装和配置 JDK 开发环境

JDK（Java Development Kit，Java 开发工具包）是 Sun 公司提供的 Java 开发环境和运行环境，是所有 Java 类的应用程序的基础。从 JDK 1.7 版本开始，由 Oracle 公司负责版本升级扩展服务。JDK 包括一组 API 和 JRE（Java 运行时环境），这些 API 是构建 Java 类应用程序的基础。

JDK 为免费开源的开发环境，任何开发人员都可以直接从官方网站下载安装程序包。本教材使用的是 JDK 1.8 版本。

1. JDK 安装步骤

（1）双击 JDK 安装程序，弹出安装对话框，如图 1-1 所示。

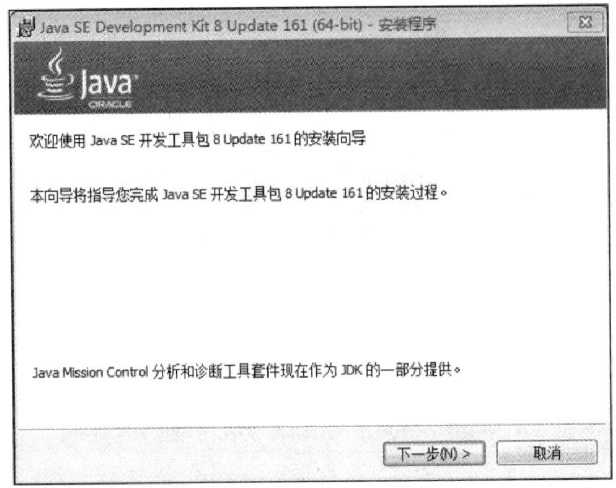

图 1-1

（2）单击“下一步”按钮，进入“定制安装”界面，如图 1-2 所示。

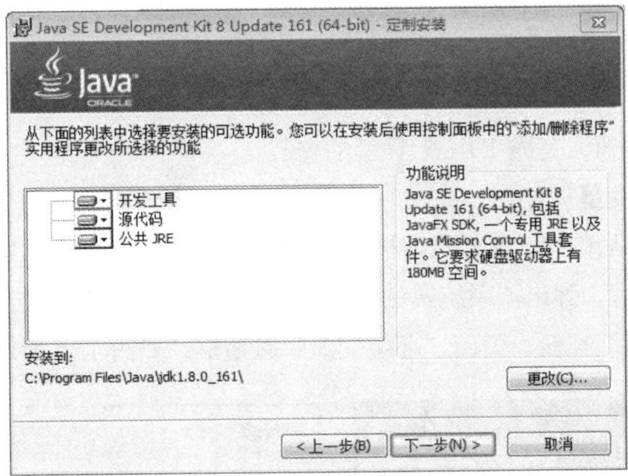

图 1-2

（3）选择安装路径。如需更换安装路径，则单击“更改”按钮，在弹出的对话框中选择安装目录的位置。注意，安装目录中不要使用中文目录名称。单击“下一步”按钮，进入正在安装界面，开始安装，如图 1-3 所示。

图 1-3

（4）安装过程中会出现 JRE 安装路径选择界面，处理方式同步骤（3）。再单击“下一步”按钮，系统进入自动安装状态，最后进入安装完成界面，如图 1-4 所示。

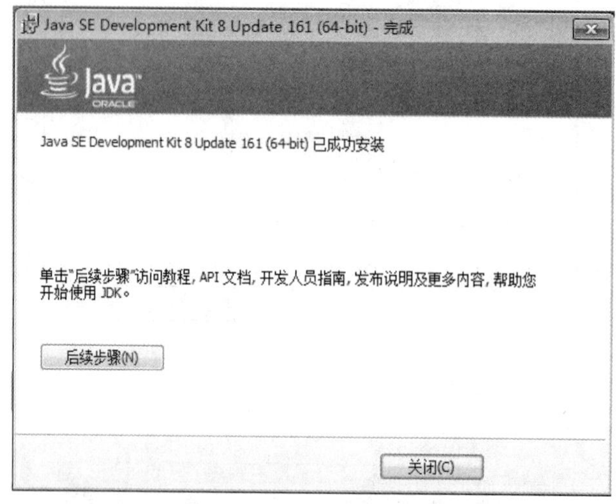

图 1-4

（5）单击“完成”按钮，完成 JDK 工具包的安装。

2. 配置 JDK 环境变量

（1）鼠标右键单击桌面的“计算机”图标，在弹出的快捷菜单中选择“属性”命令，在弹出的窗口中选择“高级系统设置”。弹出“系统属性”对话框，切换至“高级”选项卡，如图 1-5 所示。

（2）单击“环境变量”按钮，弹出“环境变量”对话框，如图 1-6 所示。

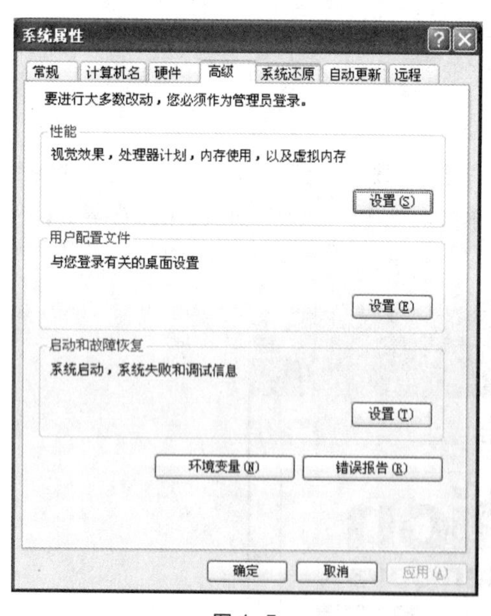

图 1-5

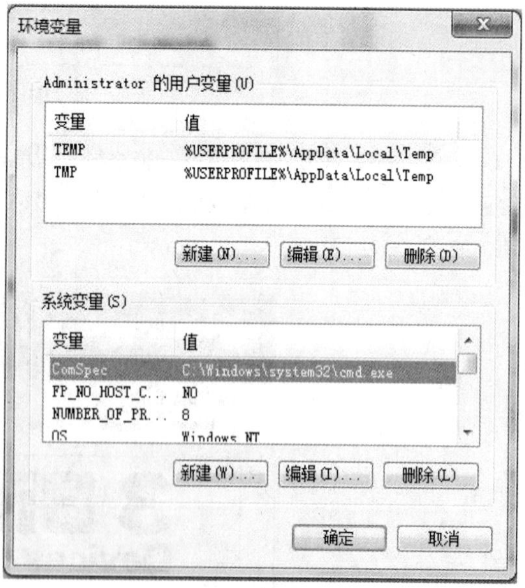

图 1-6

（3）在“环境变量”对话框的“系统变量”选项组中单击“新建”按钮，弹出“新建系统变量”对话框。在“变量名”文本框中输入“JAVA_HOME”，在“变量值”文本框中输入“C:\Program

Files\Java\jdk1.8.0_25”（JDK 的安装路径），如图 1-7 所示。单击“确定”按钮，完成设置，返回到“环境变量”对话框。

（4）在“环境变量”对话框的“系统变量”选项组中选择 Path 选项，单击“编辑”按钮，弹出“编辑系统变量”对话框。保留“变量值”文本框中的原有内容，在最后加入“;%JAVA_HOME%\bin;%JAVA_HOME%\jre\bin”，如图 1-8 所示。单击“确定”按钮，完成设置，返回到“环境变量”对话框。

图 1-7

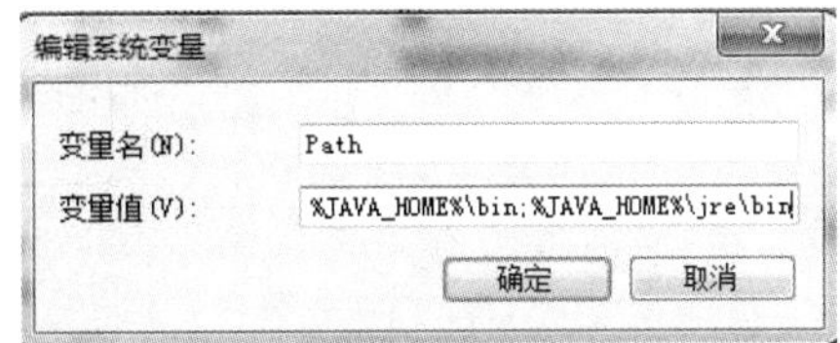

图 1-8

（5）在“环境变量”对话框中，再次单击“新建”按钮，弹出“新建系统变量”对话框。在“变量名”文本框中输入“CLASSPATH”，在“变量值”文本框中输入“.;%JAVA_HOME%\lib;%JAVA_HOME%\lib\tools.jar”，如图 1-9 所示。单击“确定”按钮，完成设置，返回到“环境变量”对话框。

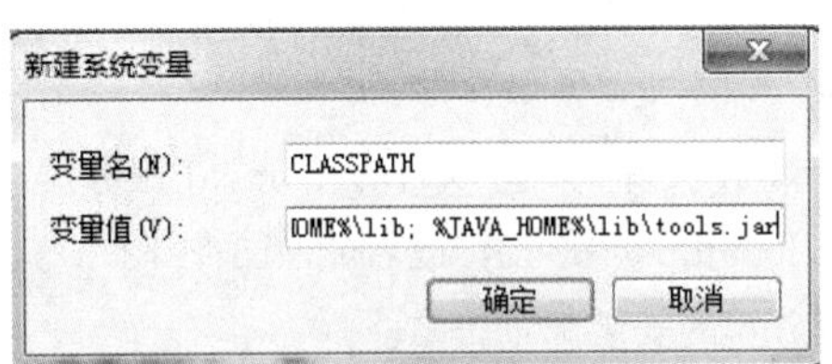

图 1-9

（6）在“环境变量”对话框中单击“确定”按钮，返回到“系统属性”对话框。在“系统属性”对话框中单击“确定”按钮，退出该对话框，完成环境变量的配置。

（二）安装 Tomcat 服务器

Tomcat 是 Apache 组织旗下的 Jakarta 项目组开发的产品，具有免费和跨平台等诸多特性。Tomcat 服务器运行稳定、性能可靠，是当今使用最广泛的 Servlet/JSP 服务器。Tomcat 已经成为学习 JSP 技术和开发中小型 Java Web 应用的首选。

1. Tomcat 的安装

Tomcat 为免费开源的产品，任何开发人员都可以直接从官方网站“http://tomcat.apache.org/”下载安装文件。本教材使用的是 Tomcat 8.0 版本。

下载的 Tomcat 为免安装版，是一个压缩包文件（apache-tomcat-8.5.31-windows-x64.zip）。将其解压缩到本地磁盘即可使用。如读者下载的是最新版本，本书内容一般来说仍然适用。

2. Tomcat 目录结构

Tomcat 服务器文件解压缩成功后将会出现 7 个文件夹，如图 1-10 所示。

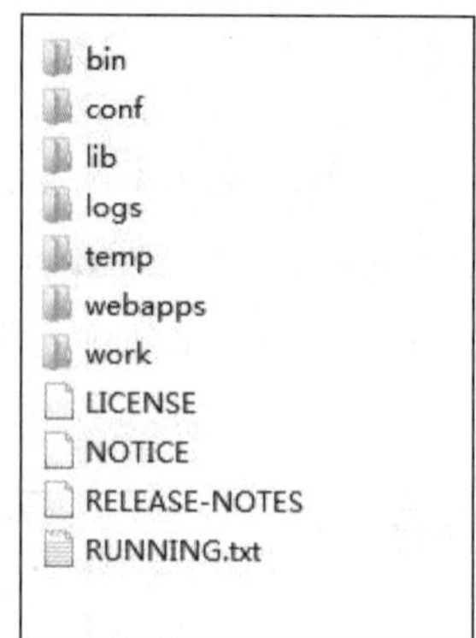

图 1-10

（1）bin 目录：存放启动、停止服务器的脚本文件。

（2）conf 目录：存放服务器的配置文件。

（3）lib 目录：存放服务器和所有的 Web 应用程序都可以访问的 JAR 包文件。

（4）logs 目录：存放服务器的日志文件。

（5）temp 目录：存放 Tomcat 运行时的临时文件。

（6）webapps 目录：Tomcat 默认的 Web 应用的发布目录。

（7）work 目录：默认情况下存放编译 JSP 文件后生成的 servlet 类文件。

3. 启动 Tomcat 服务器

执行 bin 目录下的 startup.bat 文件，就可以启动 Tomcat 服务器。Tomcat 服务器启动后，在浏览器的地址栏中输入“http://localhost:8080”或者“http://127.0.0.1:8080”，如出现 Tomcat 的测试页面，则表示 Tomcat 服务器的安装配置正常。

Tomcat 服务器默认占用 8080 端口，如果该端口已经被占用，则服务器无法正常启动。可以修改端口，位置在 conf 目录下的 servlet.xml 配置文件中。

任务四　集成开发环境的安装

任务要求

本任务要求了解集成开发环境的作用，了解 Eclipse 集成开发工具的安装，了解 IDEA 集成开发工具的安装。

任务实现

（一）集成开发工具简介

集成开发环境（Integrated Development Environment，IDE）是用于提供程序开发环境的应用程序，一般包括代码编辑器、编译器、调试器和图形用户界面工具。这是集代码编写功能、分析功能、编译功能、调试功能等于一体的开发软件服务套件。所有具备这一特性的软件或者软件套（组）件都可以称为集成开发环境。

从 JSP 诞生至今，为它量身定做的开发编译平台已不下几十种。除了 Sun 公司自身以外，还有许多的软件开发商加入到其中。下面介绍几款最为常用的 Java Web 集成开发工具。

1. NetBeans

Sun 公司推出的 NetBeans 平台是开放源码的 Java 集成开发环境，能够对 Java 应用系统的编码、编译、调试与部署提供全功能支持，并将版本控制和 XML（可扩展标记语言）编辑融入它众多的功能之中。NetBeans 的最大优势在于：它不仅能够开发各种桌面应用系统，而且能够很好地支持 Web 应用开发，支持基于 J2ME 的移动设备应用开发。

2. Eclipse

2001 年 11 月，IBM、Borland、Red Hat 等多家软件公司成立了 Eclipse.org 联盟，IBM 公司向该联盟捐赠并移交了 Eclipse 的源代码，由该联盟继续推动 Eclipse 的后续研发与更新。与商业软件不同，Eclipse 是一个完全免费的、开放源代码的、可扩展的 Java 集成开发环境，它源自于 IBM 公司耗资近 4000 万美元的一个研究项目。目前 Eclipse 得到 IBM 软件巨头及众多软件技术人员的倾力支持，极有发展前途。

3. MyEclipse

MyEclipse 企业级工作平台是对 EclipseIDE 的扩展，是在 Eclipse 基础上加上自己的插件开发而形成的功能强大的企业级集成开发环境。MyEclipse 主要用于 Java、JavaEE 及移动应用的开发。MyEclipse 的功能非常强大，支持的产品也十分广泛，尤其对各种开源产品的支持，其表现相当不错。MyEclipse 包括完备的编码、调试、测试和发布功能，完整支持 HTML、Struts、JSP、CSS、JavaScript、Spring、Hibernate 等技术和产品。

4. JBuilder

Borland 公司的 JBuilder 是世界上第一个实现跨平台的 Java 集成开发环境，也是使用最为广泛的 Java 集成开发工具之一。它是纯 Java 语言编写的编译器，系统代码中不含任何专属代码和标记，支持最新的 Java 技术。JBuilder 秉承了 Borland 产品一贯的高度集成的开发环境、豪华美观的图形界面、优质高效的编译效率等特点，适合企业级的 Java 应用系统的开发，能够轻松胜任 EJB、Web、XML 及数据库等各类应用程序的开发与部署。

5. IntelliJ IDEA

IntelliJ IDEA 集成开发环境（以下简称 IDEA），是 JetBrains 公司的产品，是 Java 编程语言开发的集成开发环境。IDEA 在业界被公认为最好的 Java 开发工具之一，尤其在智能代码助手、代码自动提示、重构、J2EE 支持、EJB 支持、各类版本工具、JUnit、CVS 整合、代码分析、创新的 GUI 设计等方面的功能可以说是超常的。

（二）Eclipse 的安装与配置

Eclipse 是一个开放源代码的、基于 Java 的可扩展集成开发平台。Eclipse 本身只是一个框架和一组服务，用于通过插件/组件构建开发环境。只要有合适的组件，Eclipse 不但能够支持开发 Java 应用程序，而且能够支持 J2EE 企业级 Java Web 开发。

Eclipse 是一款完全免费的工具，包含 Eclipse 平台、Java 开发工具、插件开发环境等内容，可以从 Eclipse 官方网站下载。对于 Java Web 程序开发来说，要下载 Eclipse for J2EE 版本。

在安装 Eclipse 之前，先要安装、配置好 JDK 和 Tomcat。

1．安装、启动 Eclipse

Eclipse IDB for Java EE 安装文件是一个压缩文件，将该文件直接解压缩到指定的安装目录下即完成了安装工作。

双击安装目录下的 eclipse.exe 文件，Eclipse 就开始运行。首次启动 Eclipse 时，会提示选择一个工作区，以便可以将相关的项目文件保存在这个工作区中。在此可以输入一个工作区保存路径位置（如 C:\Mywork）。单击“确定”按钮后，Eclipse 会出现一个欢迎界面。关掉欢迎界面，Eclipse 便进入如图 1-11 所示的工作界面。

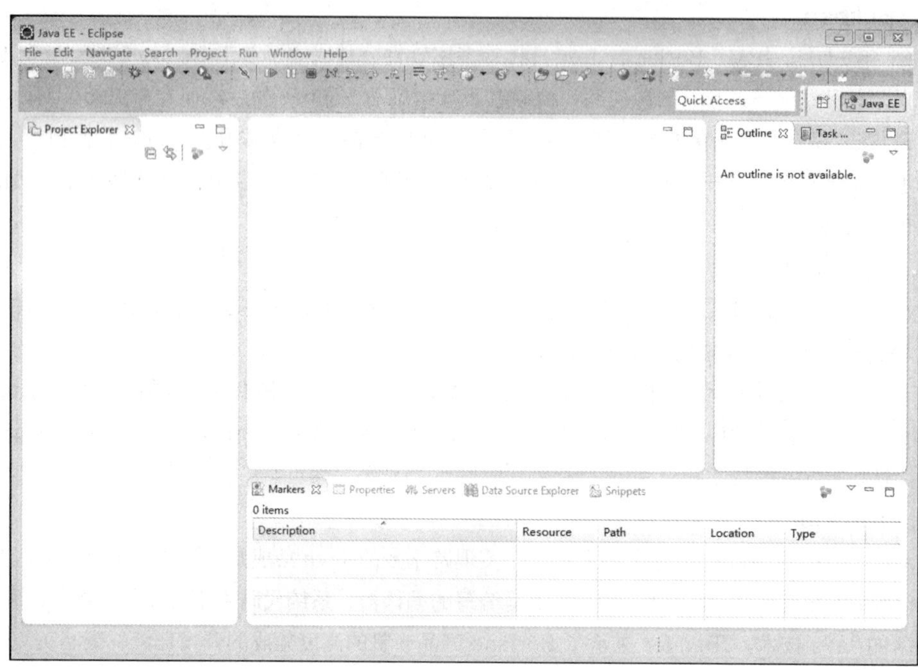

图 1-11

2．配置 Eclipse 环境

➢ 配置 JDK 开发环境

在 Eclipse 工作界面中，选择 Window（窗口）→Preferences（首选项）命令，打开 Preferences 对话框。展开对话框左侧树形列表框内的 Java 节点，选择该节点下的 Installed JREs（已安装的 JRE）子节点，对话框右侧出现如图 1-12 所示的 Installed JREs 列表框。检查列表框中 JRE 的名称、位置与所安装的 JRE 是否一致。如果不一致，修改列表框中的内容；如果一致，单击 OK 按钮。

➢ 配置 Tomcat 服务器

在发布和运行 Java Web 项目前，需要先配置 Web 服务器。本书使用的 Web 服务器是 Tomcat 服务器。

在 Eclipse 工作界面中，选择 Window（窗口）→Preferences（首选项）命令，打开 Preferences 对话框。展开对话框左侧树形列表框内的 Tomcat 节点，选择该节点下的 Advanced 子节点，在对话框右侧选择 Tomcat 的安装目录，单击 OK 按钮，如图 1-1 3 所示。

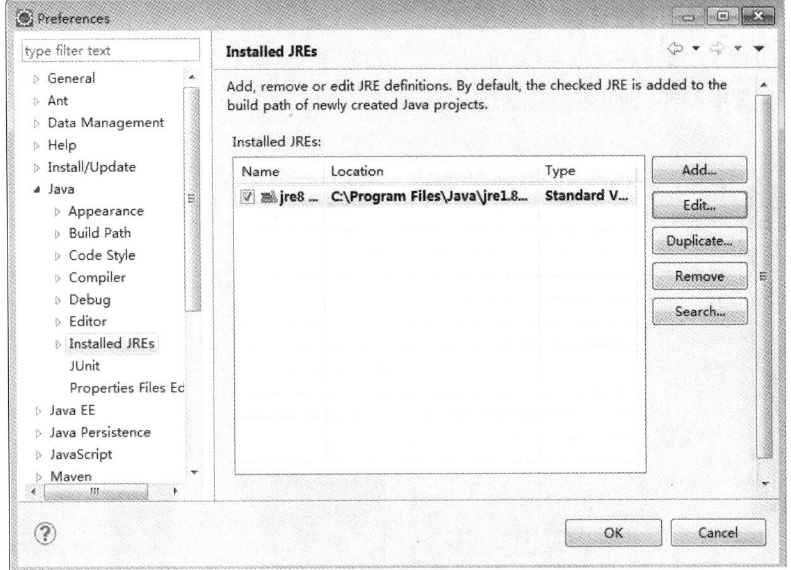

图 1-12

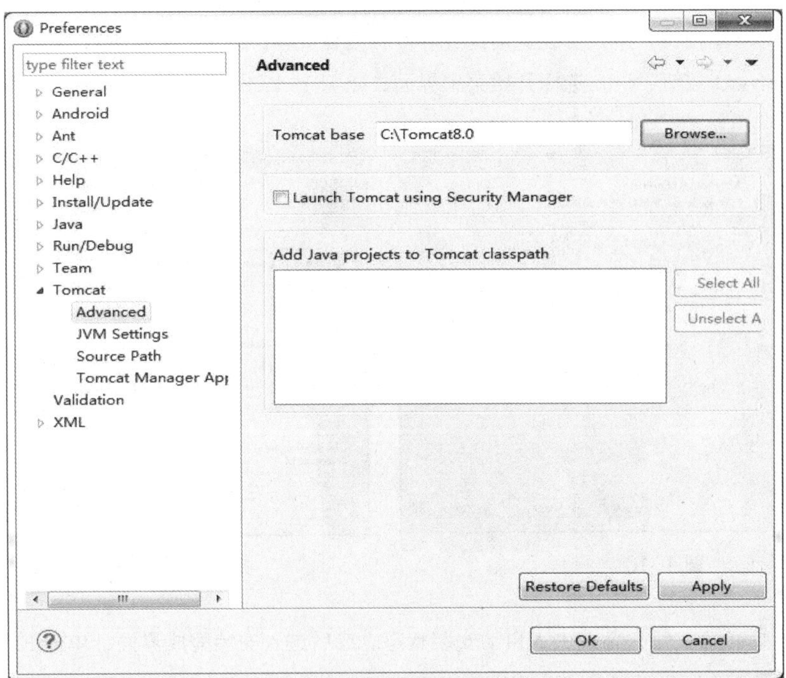

图 1-13

（三）IDEA 的安装与使用

本书使用 IntelliJ IDEA 2018 集成开发环境（以下简称 IDEA）实现 Java Web 网站的开发。可以从 IDEA 官方网站下载安装程序文件“idealU-2018.1.1.exe”。

在安装 IDEA 之前，要先安装、配置好 JDK 和 Tomcat，然后就可以开始安装 IDEA 开发环境了。

1. 安装 IDEA 集成开发环境

（1）双击 IDEA 安装程序，弹出安装对话框，如图 1-14 所示。单击 Next 按钮，进入安装位置选择界面，指定安装路径，如图 1-15 所示。

图 1-14

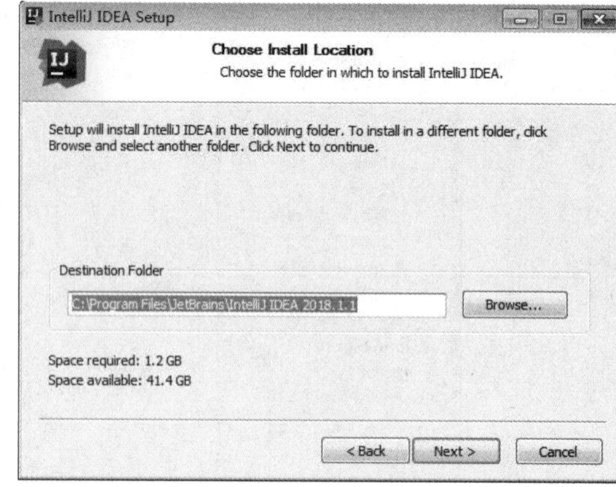

图 1-15

（2）单击 Next 按钮，进入安装选项设置界面，设置选项，如图 1-16 所示。

（3）单击 Next 按钮，进入选择开始菜单界面，如图 1-17 所示。

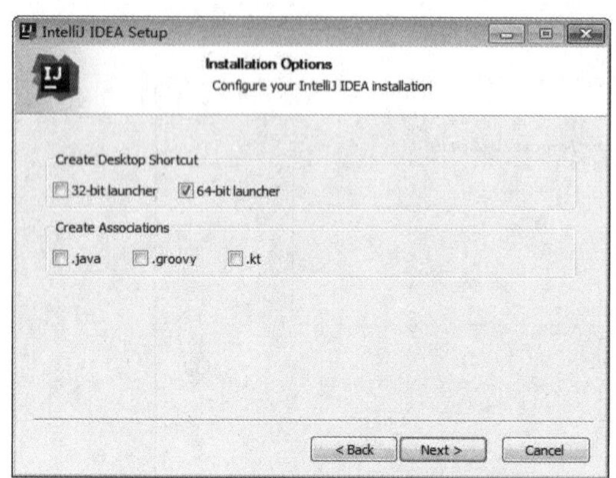

图 1-16

图 1-17

（4）单击 Install 按钮，系统进入自动安装状态。最后进入安装完成界面，单击 Finish 按钮，完成 IDEA 的安装。

2. 启动 IDEA 集成开发环境

首次启动 IDEA 集成开发环境时，需要进行官网的注册认证。

（1）双击 IDEA 程序的开始菜单启动项，弹出导入 IDEA 设置对话框，如图 1-18 所示。单击 OK 按钮，进入用户使用协议界面，单击 Accept 按钮，如图 1-19 所示。

（2）进入用户注册码填写界面，选择注册方式，如图 1-20 所示。可以通过官方网站申请获得用户注册码。

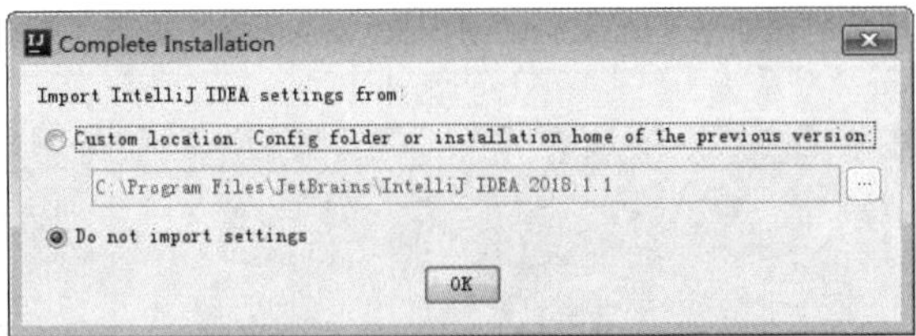

图 1-18

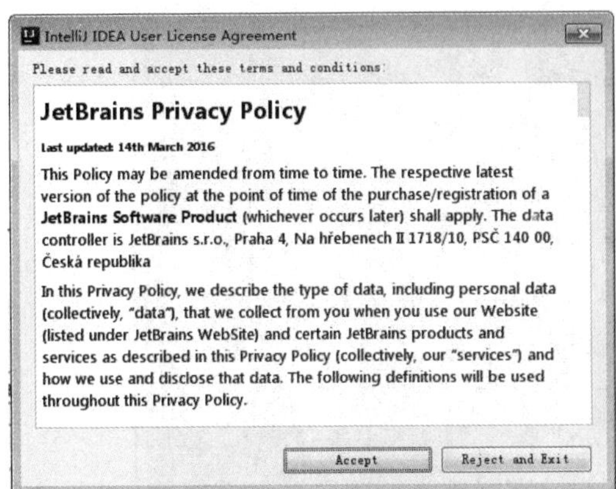

图 1-19

图 1-20

（3）单击 OK 按钮，进入自定义 IDEA 界面，如图 1-21 和图 1-22 所示。根据需要，进行设置操作。

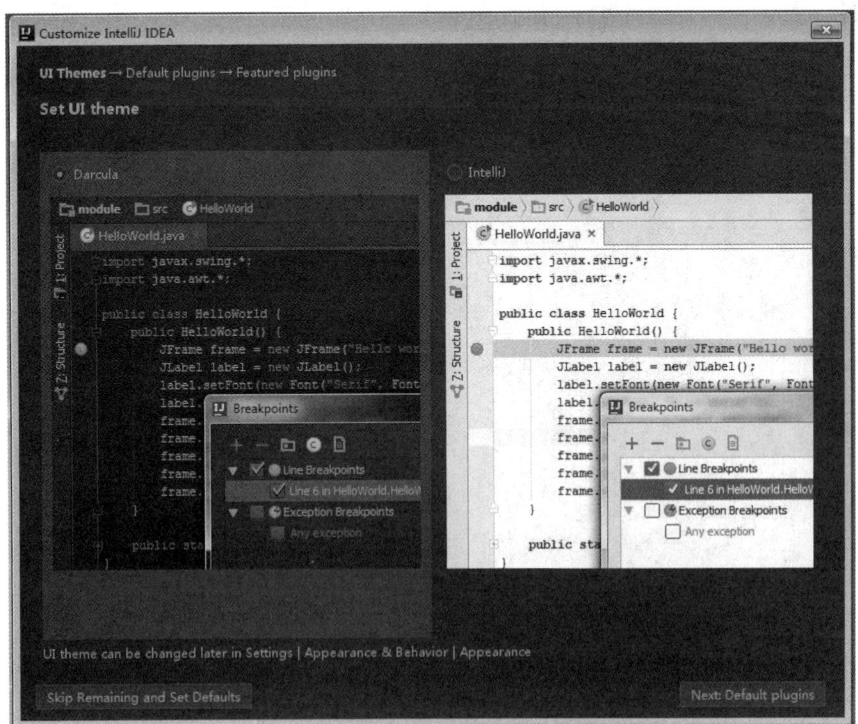

图 1-21

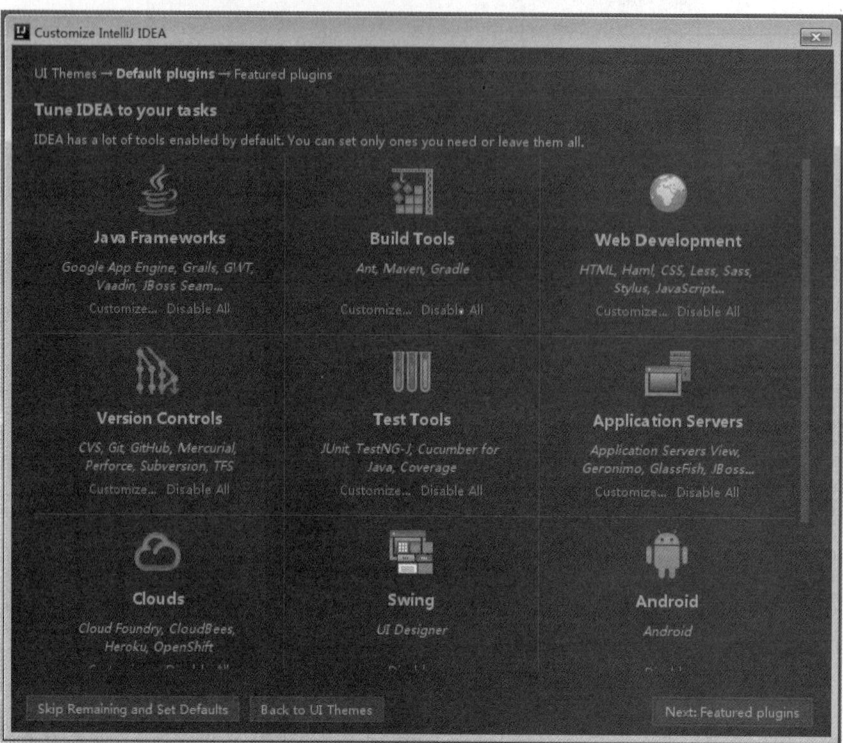

图 1-22

（4）最后进入 IDEA 的使用界面，如图 1-23 和图 1-24 所示。

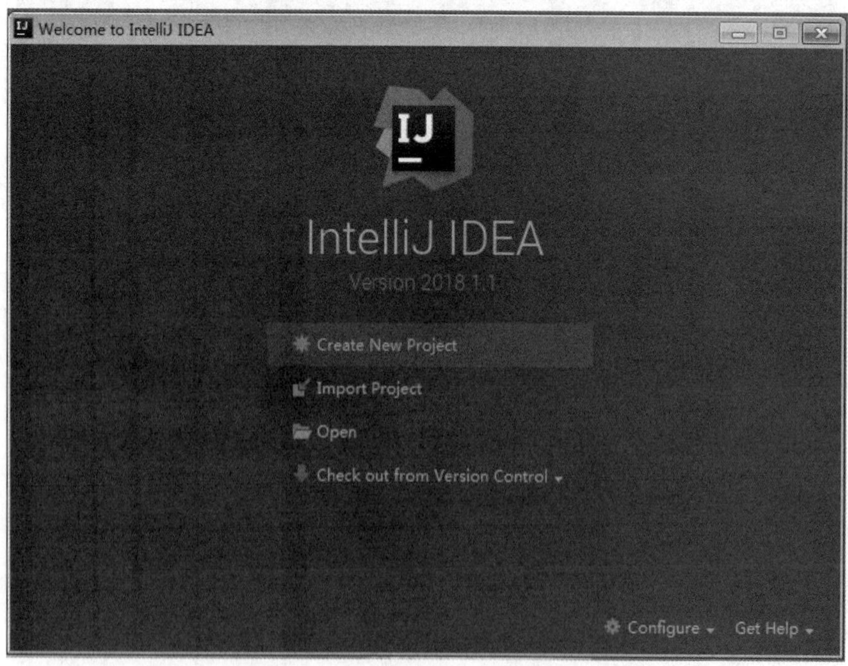

图 1-23

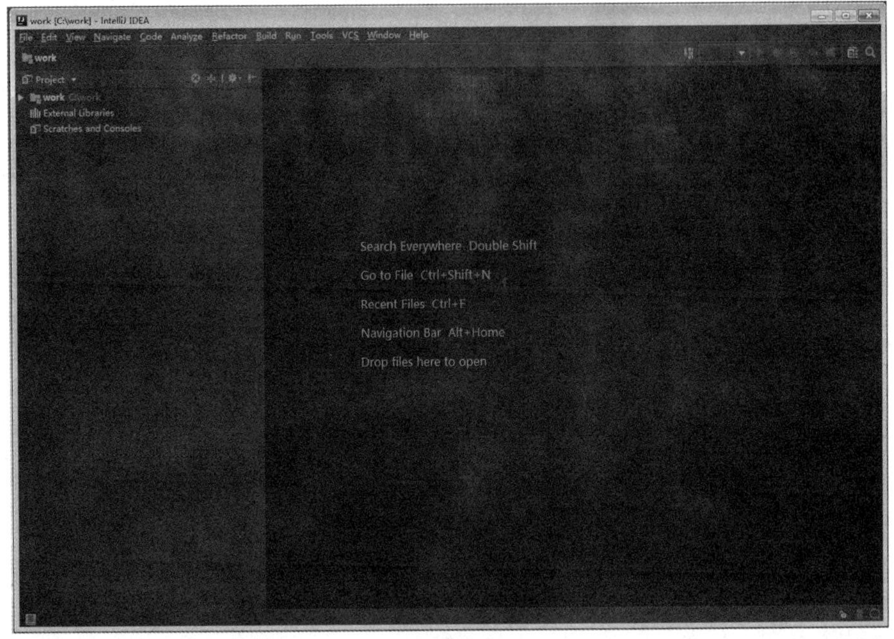

图 1-24

IDEA 集成开发环境的使用及 IDEA 环境下 Java Web 运行服务的配置，请详见本书“项目七‘天码行空’企业网站的设计与实现”的相关部分。

本章小结

本章主要介绍了现有 Web 开发技术的种类，主要的客户端技术和服务器端技术，JSP 技术的主要结构、如何搭建和配置 JSP 开发环境和如何安装集成开发环境等知识。通过本章的学习，用户对 JSP 开发技术有了一个初步的了解，并能完成开发环境的搭建，为后面的学习奠定基础。

课后练习

1．简答题

（1）简述 Web 客户端开发技术。

（2）简述 Web 服务器端开发技术。

（3）简述 JSP 页面的组成。

2．操作题

（1）下载 JDK 开发环境，安装并配置环境变量。

（2）下载 Tomcat 服务器，安装 Tomcat 服务器。

（3）下载 Eclipse for Java EE 集成开发工具，安装并完成配置。

（4）下载 IDEA 集成开发工具，安装并完成配置。

PART 02

项目二

JSP基本语法

JSP 页面由 HTML 静态元素和 JSP 动态元素两部分组成。本章主要讲解 JSP 页面的基本结构，并介绍 JSP 的基本语法。

课堂学习目标

- 理解 JSP 页面的基本结构
- 掌握 JSP 的基本语法
- 熟练使用 JSP 开发环境

素养拓展

- 培养认真和细心的编程习惯

素养拓展

任务一　JSP 页面的基本构成

任务要求

本任务要求深入理解 JSP 页面的基本构成。

任务实现

JSP 页面遵循 Java 的语法规则，是 HTML 与 Java 语言两者的融合。简言之，在静态页面中按照语法嵌入动态代码就构成了 JSP 动态页面。一个完整的 JSP 页面的构成如图 2-1 所示。

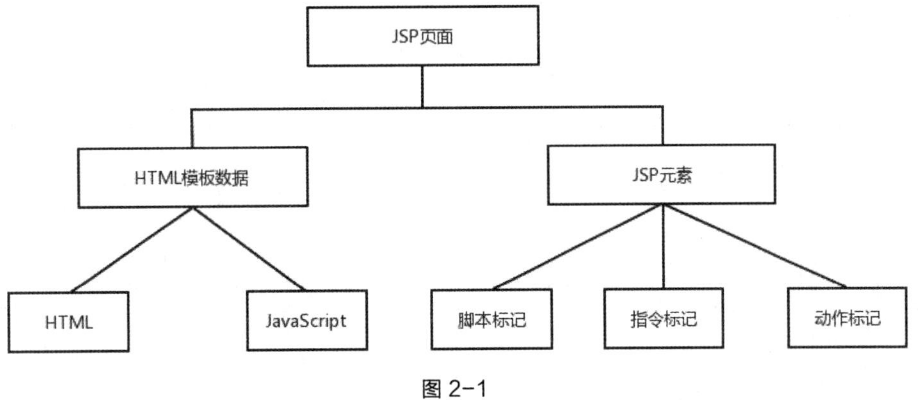

图 2-1

任务二　JSP 语法基础

任务要求

本任务要求熟练掌握 JSP 的基本语法，并能够运用脚本标记、指令标记、动作标记等 JSP 元素。

任务实现

（一）JSP 脚本标记

1. JSP 脚本段

JSP 脚本段（Scriptlet）是指一个有效的程序段，在这个程序段中可以声明要用到的变量和方法、编写 Java 语句，以及使用任何隐含的对象等。

JSP 脚本段的基本语法为<%Java 程序段%>，例如：

```
<%
int sum=0;
for( int i=0; i<=10; i++){
    sum+=I;
```

```
}
out.println("<h1>sum="+sum+"</h1>");
%>
```

➢ JSP 脚本段中只能出现 Java 代码，不能出现其他模板元素。

➢ JSP 脚本段中，Java 程序段必须严格遵循 Java 语法。

➢ 在一个 JSP 页面中可以有多个脚本段，在两个或多个脚本段间可以嵌入文本、HTML 标记和其他 JSP 元素。

➢ 多个脚本段中的代码可以相互访问，单个脚本段中的 Java 语句可以是不完整的，但多个脚本段代码组合后的结果必须是完整的，如例 2-1。

【例 2-1】 多个脚本段的相互访问。

在项目中创建 scriptlet.jsp，代码如下：

```
<%@ page contentType="text/html;charset=GB2312" language="java" %>
<html>
<head>
    <title>多个脚本段的相互访问</title>
</head>
<body>
<%!
    int i;
%>
<%
    for(i=0; i<=10; i++)
    {
%>
<h1>你好！张先生</h1>
<%
    }
%>
</body>
</html>
```

在该页面中 for 循环被拆分到两个脚本段，中间插入一段 HTML 语句，这两个脚本段相互访问，构成完整的循环，效果如图 2-2 所示。

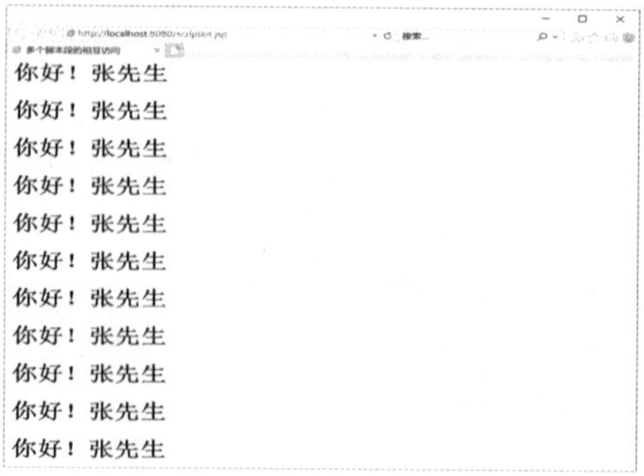

你好！张先生
你好！张先生
你好！张先生
你好！张先生
你好！张先生
你好！张先生
你好！张先生
你好！张先生
你好！张先生
你好！张先生
你好！张先生

图 2-2

2. JSP 声明

在 JSP 页面中，可以声明合法的变量和方法，变量类型可以是 Java 语言允许的任何数据类型。这种声明是全局变量。

JSP 声明（Declaration）的基本语法为<%! 声明 1; 声明 2; ... 声明 n; %>，声明的本质就是将声明的变量和方法作为 Servlet 类的变量和方法。下面用一个例子来解释如何声明变量和方法。

【例 2-2】 声明变量和方法。

本例在 declaration.jsp 中声明一个整型变量和一个方法，并在后面的代码段中加以调用。代码如下：

```
<%@ page contentType="text/html;charset=GB2312" language="java" %>
<html>
<head>
    <title>声明变量和方法</title>
</head>
<%!
    //声明一个整型变量
    public int count;

    //声明一个方法
    public String info() {
        return "hello";
    }
%>
<body>
<%
    //将count的值输出后再加1
    out.println(count++);
%>
<br/>
<%
    //输出info()方法的返回值
    out.println(info());
%>
</body>
</html>
```

declaration.jsp 运行后的效果如图 2-3 所示，每次刷新后，count 变量都会自加 1，如图 2-4 所示。

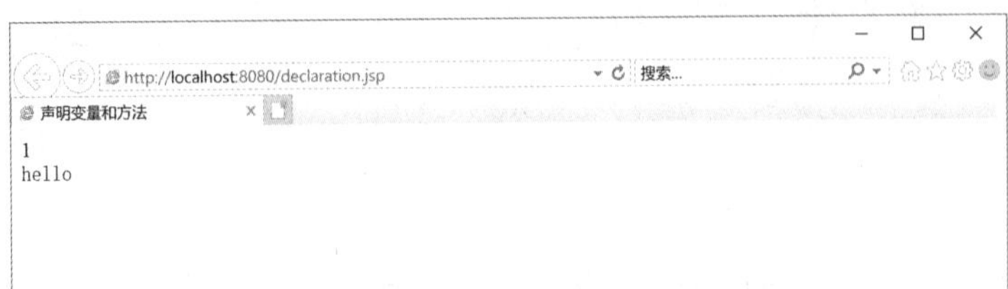

图 2-3

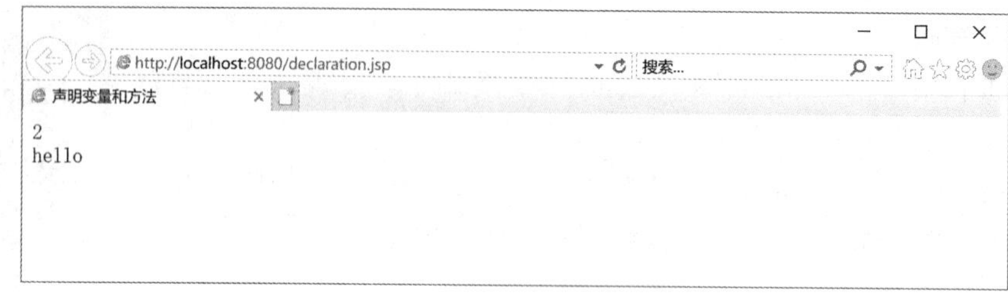

图 2-4

➢ 声明必须以分号（；）结尾。

➢ 可以直接使用在<%@ page %>中已经声明的变量和方法，不需要对它们重新进行声明。

➢ 一个声明仅在一个页面中生效。如果要在多个页面中用到该声明，则可将它们写成一个单独的文件，然后用<%@ include %>或<jsp:include>包含进来。

3. JSP 表达式

在 JSP 页面中，可以用表达式（Expression）将程序数据输出到客户端，其等价于 out.print。表达式元素表示的是一个在脚本语言中定义的表达式，在运行后会自动转换为字符串，然后插入到这个表达式在 JSP 页面中的位置并显示。

JSP 表达式的基本语法为<%=变量或表达式%>，表达式的本质就是将 JSP 页面转化为 Servlet 后，使用 out.print()将表达式的值输出。下面用一个例子来理解表达式的妙用。

【例 2-3】 表达式实例。

本例在 expression.jsp 中定义了字符串变量 url，并利用表达式指定了超级链接的页面及处理表单信息的页面。代码如下：

```
<%@ page contentType="text/html;charset=GB2312" language="java" %>
<html>
<head>
    <title>表达式实例1</title>
</head>
<%!
    String url="expressionHref.jsp";
%>
<body>
<a href="<%=url%>">点击跳转</a>    //超级链接指向url指代的页面

<form action="<%=url%>">   //url指代的页面规定了处理表单信息的页面
    <input type="submit" value="点击跳转"/>
</form>
</body>
</html>
```

创建 expressionHref.jsp，代码如下：

```
<%@ page contentType="text/html;charset=GB2312" language="java" %>
<html>
```

```
<head>
    <title>表达式实例2</title>
</head>
<body>
<h1>表达式实例</h1>
</body>
</html>
```

页面 expression.jsp 的运行效果如图 2-5 所示，单击超级链接“点击跳转”或单击按钮“点击跳转”，都会定向到 expressionHref.jsp，如图 2-6 所示。

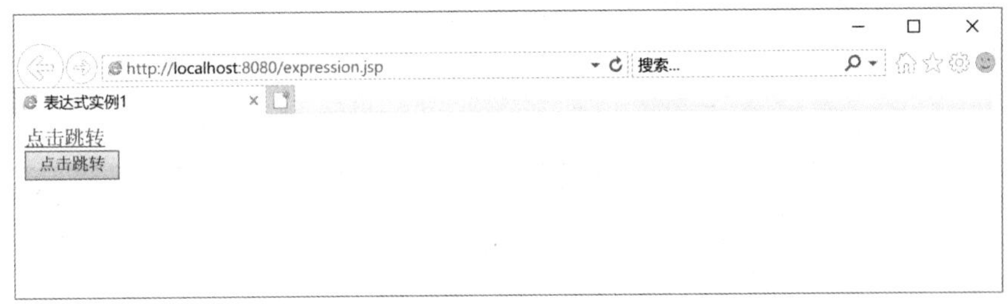

图 2-5

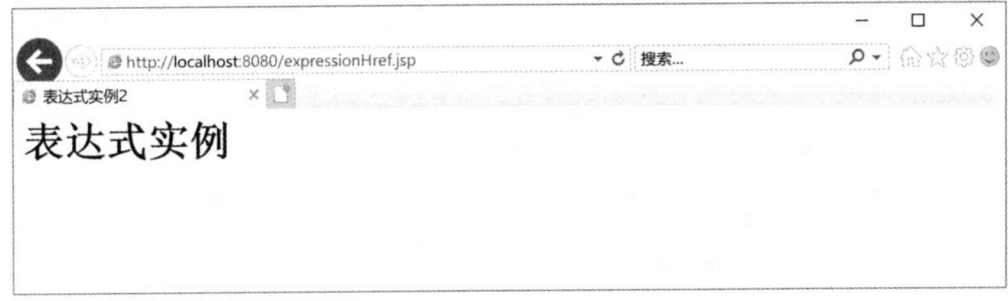

图 2-6

JSP 表达式不能用分号（;）作为结束符。

4. JSP 注释

在 JSP 页面中，注释分为两大类：静态注释和动态注释。静态注释是直接使用 HTML 风格的注释，这类注释在浏览器中查看源文件时是可以看到注释内容的；动态注释包括 Java 注释和 JSP 注释两种，这类注释在浏览器中查看源文件时是看不到注释内容的。

注释的语法为：

➢ 静态注释

```
<!-- 注释内容 -->
```

➢ Java 注释

```
//单行注释
/*
  多行注释
*/
```

➢ JSP 注释

```
<%-- 注释内容 --%>
```

（二）JSP 指令标记

JSP 指令标记（Directive Element）是为 JSP 引擎设计的，该类标记并不直接产生任何可见的输出，而是告诉 JSP 引擎如何处理 JSP 页面的其余部分。例如，可以指定一个专门的错误处理网页，当 JSP 页面出现错误时，可以由 JSP 引擎自动地调用错误处理网页。

常用的 JSP 指令标记包括 page 页面指令标记、include 静态包含指令标记、taglib 指令标记。

1. page 页面指令标记

page 指令用于定义与整个 JSP 页面相关的各种属性。其基本语法为：

```
<%@page 属性1="值1"  属性2="值2" ... 属性n="值n"%>
```

page 指令常用的属性和默认值如表 2-1 所示。

表 2-1 page 指令的常用属性

属性	解释	默认值
language	声明要使用的脚本语言，暂时只能支持 Java	Java
import	在程序中导入一个或多个类和包	
session	设定客户是否需要 HTTP Session，取值为 true/false 若为 true，则表示 session 是有用的	true
autoFlash	设置缓冲区填满时是否进行缓冲自动刷新	true
isThreadSafe	设置 JSP 页面是否支持多线程，值为 false 时限制每次只能有一个用户访问该页面	true
isErrorPage	指定当前页面是否可以作为另一页面的错误处理页码	false
errorPage	指定当前网页的出错处理网页的 URL	
contentType	指定 JSP 字符的编码和 JSP 页面响应的 MIME 类型，格式为"MIME 类型；字符集类型"	"text/html;charset=gb2312"

例如，在某个 JSP 页面中，如果需要导入 Java 的 sql 包，并设置错误处理页面，代码如下：

```
<%@page contentType="text/html; charset=GB2312"%>
<%@page import="java.sql.*"%>
<%@page errorPage="err.jsp"%>
```

无论 page 指令出现在 JSP 页面中的什么地方，它作用的都是整个 JSP 页面（包括静态的包含文件，但不能作用于动态的包含文件），为了保持程序的可读性和遵循良好的编程习惯，page 指令最好是放在整个 JSP 页面的起始位置。

在一个 JSP 页面中可以使用多个<%@page %>指令，但其中的属性只能用一次，不过 import 属性例外，它可以多次出现，引入多个类和包，这和 Java 中的 import 语句类似。

2. include 静态包含指令标记

include 指令用于引入其他 JSP 页面，引入后，JSP 引擎会将这两个 JSP 页面翻译成一个 servlet。include 指令通常也被称为静态引入指令。其基本语法为：

```
<%@include file="相对URL"%>
```

所谓静态，是指 file 属性的值不能是一个变量，也不可以在 file 所指定的文件后添加任何参数。

【例 2-4】 使用 include 指令标记的实例。

在 include.jsp 页面中使用 include 指令引入相同文件夹下的 head.jsp 页面，代码如下：

```
<%@ page contentType="text/html;charset=GB2312" language="java" %>
<html>
<head>
    <title>include指令标记实例</title>
    <style type="text/css">
      p{color:deepskyblue; text-align: left; size:5px;}
    </style>
</head>
<body>
<%@include file="head.jsp"%>
<p>  include指令用于引入其他JSP页面，引入后，JSP引擎会将这两个JSP页面翻译成一个servlet。
    因此include指令通常也被称为静态引入。</p>
</body>
</html>
```

创建 head.jsp，代码如下：

```
<%@ page contentType="text/html;charset=GB2312" language="java" %>
<html>
<head>
    <title>被引入的头文件</title>
    <style type="text/css">
      h1{color:red; text-align:center; size: 7px; }
    </style>
</head>
<body>
<h1>INCLUDE指令标记的使用</h1>
</body>
</html>
```

运行 include.jsp 的效果如图 2-7 所示。

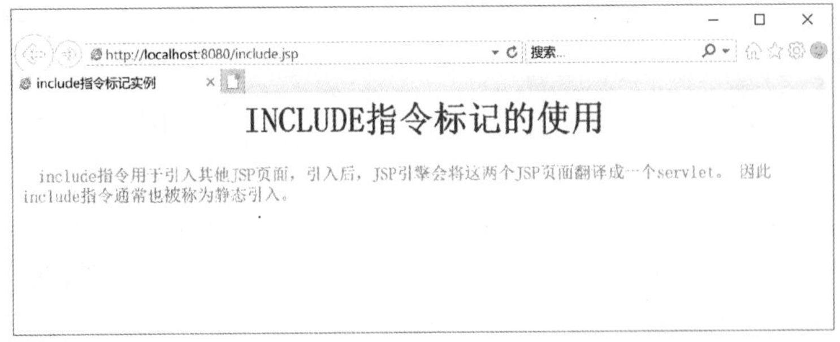

图 2-7

➢ 使用 include 指令引入的文件必须遵循 JSP 语法。

➢ 使用 include 指令引入的文件可以使用任意的扩展名，但都会被 JSP 引擎按照 JSP 页面的处理方式处理，为了见名知意，建议使用“.jspf”（JSP fragment，即 JSP 片段）作为静态引入文件的扩展名。

➢ 使用 include 指令引入的文件中不要包含<html></html>、<body></body>等标记，因为这样会影响引入文件中同样的标记。

➢ 引入和被引入文件中的指令不能冲突（page 指令中的 pageEncoding 和 import 属性除外）。

3. taglib 指令标记

taglib 指令用于引入 JSP 页面中需要使用的标签库的定义，以便在页面中使用标签库定义的标签。其基本语法为：

```
<%@taglib uri="标签库URI"  prefix="自定义标签的前缀"%>
```

属性 uri（Uniform Resource Identifier，统一资源标识符）用来唯一确定标签库的路径，属性 prefix 定义了使用此标签库的前缀。

例如：

```
<%@taglib uri="http://www.tjdz.net/tags" prefix="public"%>
<public:loop>
...
</public:loop>
```

定义标签时，不能使用 jsp、jspx、java、javax、servlet、sun、sunw 作为前缀，这些前缀是 JSP 保留的。

（三）JSP 动作标记

与 JSP 指令标记不同，JSP 动作标记（Action Element）是在客户端请求时动态执行的。JSP 动作标记是一种特殊标签，并且以前缀 jsp 和其他的 HTML 标签相区别。利用 JSP 动作标记可以实现很多功能，包括动态地插入文件、重用 JavaBean 组件、把用户重定向到另外的页面、为 Java 插件生成 HTML 代码等。

微课：Include 动作标记

1. <jsp:include>

<jsp:include>动作标记可以用来包含其他静态或动态文件。其基本语法为：

➢ 不带参数

```
<jsp:include page="相对URL" flush="true|false"/>
```

➢ 带参数

```
<jsp:include page="相对URL" flush="true|false">
<jsp:param name="属性名" value="属性值"/>
<jsp:param...>
</jsp:include>
```

其中，属性 page 指向的是被包含文件的相对路径；当属性 flush 为 true 时，表示实时输出缓冲区，它的默认值是 false。<jsp:param>子句能传递一个或多个参数给动态文件，也可以在一个页面中使用多个<jsp:param>来传递多个参数给动态文件。

【例 2-5】 带参数的 inlcude 动作标记。

本例在 includeJsp.jsp 中通过动作标记<jsp:include>引入 date.jsp，并利用<jsp:param>向被引入的页面传递参数。此处要通过 request 内置对象的 setCharacterEncoding("编码格式")来设定编码格式，以避免传递中文时产生乱码。includeJsp.jsp 的代码如下：

```
<%@ page language="java" contentType="text/html; charset=UTF-8"
         pageEncoding="UTF-8"%>
```

```
<html>
<head>
    <meta http-equiv="Content-Type" content="text/html; charset=UTF-8">
    <title>include动作标记</title>
</head>
<body>
<h1>include动作标记</h1>
<hr>
<%
    request.setCharacterEncoding("UTF-8");//设定request的格式编码，防止传递中文时产生乱码
%>
<jsp:include page="date.jsp">
    <jsp:param name="username" value="张先生"/>
</jsp:include>
</body>
</html>
```

创建 date.jsp，代码如下：

```
<%@ page language="java" contentType="text/html; charset=UTF-8"
        pageEncoding="UTF-8"%>
<%@page import="java.text.SimpleDateFormat"%>
<%@page import="java.util.Date"%>
<html>
<head>
    <meta http-equiv="Content-Type" content="text/html; charset=UTF-8">
    <title>date页面</title>
</head>
<body>
<%
    String name=request.getParameter("username");
    out.println("你好，"+name+"!今天的日期是：");
    Date date = new Date();
    SimpleDateFormat sdf = new SimpleDateFormat("yyyy年MM月dd日");

    String string = sdf.format(date);
    out.println(string);
%>
</body>
</html>
```

运行 includeJsp.jsp 的效果如图 2-8 所示。

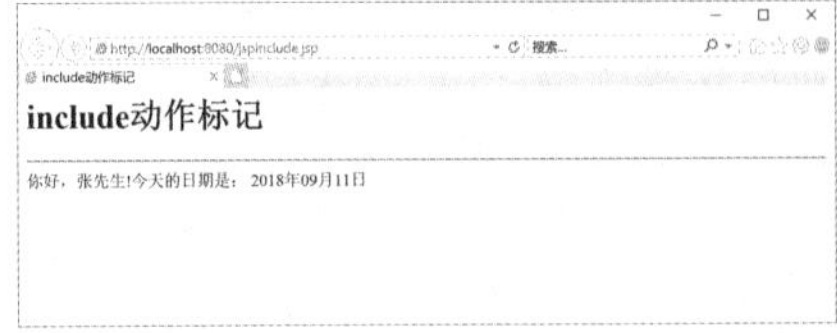

图 2-8

指令标签 include 和动作标签 include 的异同：指令标签 inlcude 是将静态嵌入的文件作为主文件的一部分，所以主文件和子文件其实是一体的；动作标签 include 是动态嵌入文件，子文件不必考虑主文件的属性，子文件是独立的。指令标签在编译时就将子文件载入；动作标签在运行时才将子文件载入。

微课：Forward 动作标记

2. <jsp:forward>

<jsp:forward>动作标记用于在服务器端终止当前页面的运行，并重定向到其他指定页面。重定向的目标可以是静态的 HTML 页面、JSP 页面，或者是一个程序段。其基本语法为：

➢ 不带参数

```
<jsp:forward page="页面URL">
```

➢ 带参数

```
<jsp:forward page="页面URL">
 <jsp:param name="属性名" value="属性值"/>
 ...
</jsp:forward>
```

其中，属性 page 指向的是重定向的页面路径。

【例 2-6】 带参数的 forward 动作标记。

本例在 forward.jsp 中通过<jsp:forward>动作标签将页面重定向到 forwardTo.jsp，并利用<jsp:param>传递参数。在 forwardTo.jsp 中，利用 request 内置对象中的 getParameter("变量名")来接收传递过来的参数。创建 forward.jsp，代码如下：

```
<%@ page contentType="text/html;charset=UTF-8" language="java" %>
<html>
<head>
    <title>带参数的forward动作标记</title>
</head>
<body>
<%!
    int i=0;
%>
<jsp:forward page="forwardTo.jsp">
    <jsp:param name="username" value="MrZhang"/>
    <jsp:param name="password" value="abc123"/>
</jsp:forward>

<p>这里的表达式能够输出吗？<%=i%></p>
</body>
</html>
```

创建 forwardTo.jsp，代码如下：

```
<%@ page contentType="text/html;charset=UTF-8" language="java" %>
<html>
<head>
    <title>带参数的forward动作标记2</title>
```

```
</head>
<body bgcolor="#00ffff">
<%
    String name=request.getParameter("username");
    String pw=request.getParameter("password");
    out.println("您的用户名是："+name+"<br/>");
    out.println("您的密码是："+pw);
%>
</body>
</html>
```

forward.jsp 的运行效果如图 2-9 所示。

图 2-9

forward.jsp 执行到<jsp:forward>标记出现处时停止当前页面的执行，并重新定向到新的页面，也就是说，forward.jsp 中<jsp:forward>标记后的部分不执行。如图 2-9 所示，并没有输出表达式<%=i%>。

<jsp:forward>动作标记执行的是服务器端的跳转，浏览器地址不变，如图 2-9 中标方框所示，地址仍为.../forward.jsp。

3. <jsp:param>

<jsp:param>动作标记用来传递参数给 JSP 页面。其基本语法为：

```
<jsp:param name="参数名" value="{参数值|<%=表达式%>}"/>
```

其中，属性 name 表示传递的参数名称，属性 value 表示设置参数的值。

JSP 标记不同于 HTML 标记，属性值必须加上双引号，否则执行时会报错。

使用<jsp:param>动作标记来传递参数，在 JSP 页面中通过 request.getParameter("属性名")来获取参数的值。

<jsp:param>动作标记必须配合<jsp:include>、<jsp:forward>或<jsp:plugin>等标记使用，在加载外部程序或是网页转换的时候，传递参数给另一个 JSP 页面。单独使用<jsp.param>没有意义。

4. <jsp:plugin>

<jsp:plugin>动作标记用于在 JSP 网页中加载 Java Applet 或 JavaBean 程序组件，与 HTML 的<Applet>与<Object>标签有类似的功能。<jsp.plugin>动作标记的常用属性如表 2-2 所示。

表 2-2　<jsp:plugin>动作标记的常用属性

属性	解释
type	加载 Java 程序的类型，可设置的值有 Applet 和 Bean
code	加载 Java 程序编译后的类名称，如 showpic.class
codebase	编译后 Java 程序类所在的目录，可设置绝对路径或者相对路径。若未设置此属性，则以当前执行网页所在的目录为默认值
name	用来加载对 Java Applet 和 JavaBean 程序设置一个用以识别的名称
align	设置加载的程序在窗口的对齐方式，可设置的有 bottom（下对齐）、top（上对齐）、middle（中对齐）、left（左对齐）、right（右对齐）
height	加载的程序在窗口中显示的高度
width	加载的程序在窗口中显示的宽度
hspace	加载程序的显示区与网页其他内容的水平间距
vspace	加载程序的显示区与网页其他内容的垂直间距
<jsp:params>	若要传递参数给加载的程序，必须在<jsp:params>和</jsp:params>的起始标签和结束标签中使用<jsp: param>操作元素来设置

5. <jsp:useBean>、<jsp:setProperty>和<jsp:getProperty>动作标记

➢ <jsp:useBean>动作标记

<jsp:useBean>动作标记用来加载 JSP 页面中使用的 JavaBean，其语法格式如下：

```
</jsp:useBean id="JavaBean实例名称" scope="page|request|session|application" class="package.class">
< </jsp:useBean>
```

其中，id 指定该 JavaBean 实例变量的名称。scope 指定该 Bean 变量的有效范围：page 指只在当前 JSP 页面有效；request 指在任何执行相同请求的 JSP 页面中使用该 Bean，直到页面执行完毕；session 指从创建该 Bean 开始，在相同 session 下的 JSP 页面中可以使用该 Bean；application 指从创建该 Bean 开始，在相同 application 下的 JSP 页面可以使用该 Bean。

例如：

```
<jsp:useBean id="clock" scope="page" class="java.util.Date"/>
```

➢ <jsp:setProperty>动作标记

<jsp:setProperty>动作标记用于设置已经实例化的 Bean 对象的属性，其基本语法格式为：

```
<jsp:setProperty
  name="JavaBean实例名称"
  {
   property="*"|
   property="属性名" [param="参数"]|
   property="属性名" value="{String|<%=表达式%>}"
}
/>
```

<jsp:setProperty>中 name 值必须和<jsp:useBean>中的 id 值相同，且大小写敏感。

➢ <jsp:getProperty>动作标记

<jsp:getProperty>动作标记可获取 Bean 的属性值，用于在页面中显示。其基本语法格式为：

```
<jsp:getProperty name="JavaBean实例名称" property="属性名"/>
```

任务三　拓展实训

任务要求

构建一个主题网站，进一步掌握 JSP 的基础语法。

任务实现

（一）主题网站的运行效果

在浏览主题网站时会发现，这些网站的不同频道都是用相同的 Logo 和频道导航做题头。本任务就是要完成一个主题网站，如图 2-10～图 2-12 所示。如果在每个页面的题头部分都放置网站的 Logo 和频道导航，导致的问题是，当网站 Logo 或频道需要更换时，需要对每一个频道的网页重新设计，这样维护既费时又费力。因此，为了便于实现快速维护，将两个相对独立的网页拼接成如图 2-10 所示的页面，更改 Logo 图片时，只需要对 top.html 进行调整即可。

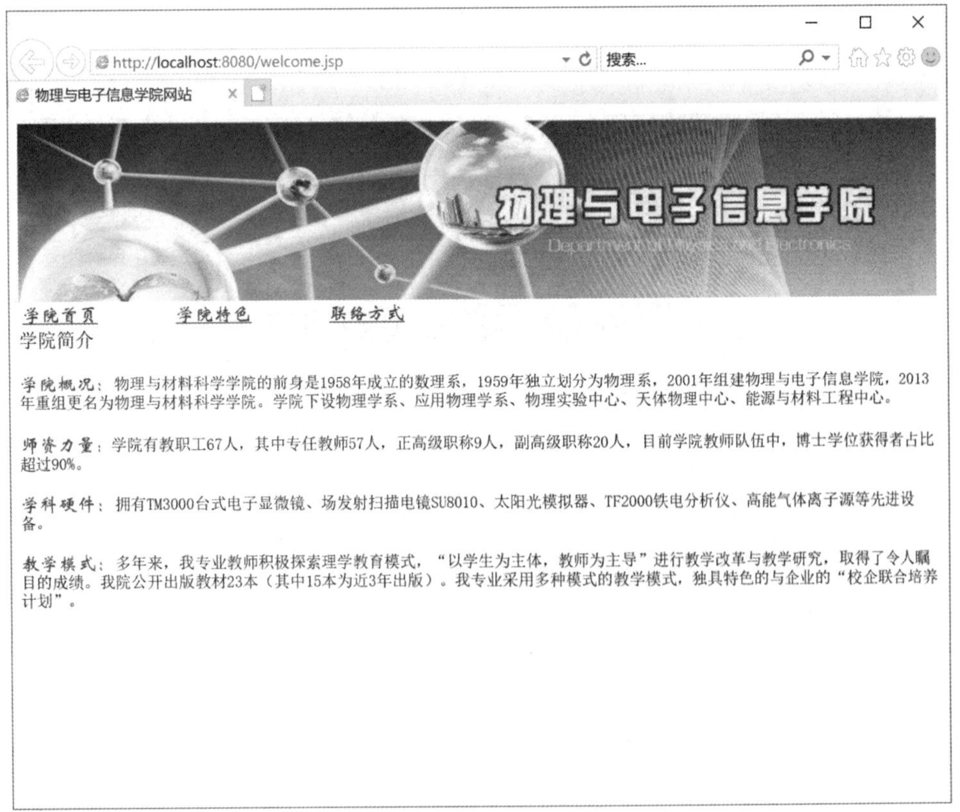

图 2-10

http://localhost:8080/index.jsp?choice=2
搜索...
物理与电子信息学院网站
物理与电子信息学院
学院首页
学院特色
联络方式
学院现有本科生739人，研究生106人。学院现有实验教学面积5000平米。在硬件条件上，经过天津市连续5个五年计划的投入、中央与地方共建项目以及学校实验室专项建设等多次成规模的投资，学院的基础实验室和专业实验室逐步完善。目前学院的教学科研仪器设备总值达到4000万元，为“强化实践育人”和高水平科学研究奠定了基础。

图 2-11

http://localhost:8080/contact.jsp
搜索...
物理与电子信息学院
物理与电子信息学院
学院首页
学院特色
联络方式
通讯地址：天津市津南区海河教园区职业技术学院
物理与电子信息学院
邮政编码：300000 办公地点： 学校行政楼507
Email：zyjsxy@tjzy.edu.cn 电话：27778777
网址：http://tjzy.edu.cn/wl/

图 2-12

（二）功能设计

1. 创建工程

启动 IntelliJ IDEA，创建 Web 应用并命名为 chapter2，在 web 文件夹下创建 images 文件夹，并将 top.jpg 导入到 images 文件夹中。在 web 文件夹下创建 JSP 类型文件 welcome.jsp、index.jsp、feature.jsp、contact.jsp 和 HTML 类型文件 top.html、welcome.html、feature.html、contact.html。

2. 将已有资源添加到当前工程

展开 chapter2 工程，可以看到 chapter2 工程下的 web 子文件夹。选中 web 子文件夹，单击鼠标右键，在弹出的快捷菜单中选择 New→Directory 命令，创建 images 文件夹，将制作好的图片文件 top.jpg 复制到该文件夹。

3. HTML 页面设计

首先创建 top.html、welcome.html、feature.html 和 contact.html 页面，如图 2-10、图 2-11、图 2-12 所示。

top.html 代码如下：

```
<%@ page pageEncoding="gb2312"%>
<html>
  <head>
    <title>top.html</title>

    <meta http-equiv="keywords" content="keyword1,keyword2,keyword3">
    <meta http-equiv="description" content="this is my page">
    <meta http-equiv="content-type" content="text/html; charset=gb2312">
  </head>
  <body>
 <img src="images/top.jpg" width="780" height="150">
 <table width="50%" border="0">
   <tr>
     <td height="19"><font color="#FF0000" face="华文新魏"><a href="index.jsp?choice=1">学院首页
</a></font></td>
     <td><font color="#FF0000" face="华文新魏"><a href="index.jsp?choice=2">学院特色</a></font>
</td>
     <td><font color="#FF0000" face="华文新魏"><a href="index.jsp?choice=3">联络方式</a></font>
</td>
 </table>
 </body>
 </html>
```

welcome.html 代码如下：

```
<%@ page pageEncoding="gb2312"%>
<html>
  <head>
    <title>welcome.html</title>
    <meta http-equiv="keywords" content="keyword1,keyword2,keyword3">
    <meta http-equiv="description" content="this is my page">
```

```
        <meta http-equiv="content-type" content="text/html; charset=gb2312">
    </head>
    <body>
    学院简介<br>
    <br>
    <font color="#FF0000" face="华文新魏">学院概况：</font><font size="2">物理与材料科学学院的前身是1958年成立的数理系，1959年独立划分为物理系，2001年组建物理与电子信息学院，2013年重组更名为物理与材料科学学院。学院下设物理学系、应用物理学系、物理实验中心、天体物理中心、能源与材料工程中心。</font><br>
    <br>
    <font color="#FF0000" face="华文新魏">师资力量<font size="2">：</font></font><font size="2">学院有教职工67人，其中专任教师57人，正高级职称9人，副高级职称20人，目前学院教师队伍中，博士学位获得者占比超过90%。</font><br>
    <font color="#FF0000" face="华文新魏"><br>
    学科硬件：</font><font size="2">拥有TM3000台式电子显微镜、场发射扫描电镜SU8010、太阳光模拟器、TF2000铁电分析仪、高能气体离子源等先进设备。</font><br>
    <font color="#FF0000" face="华文新魏"><br>
    教学模式：</font><font size="2">多年来，我专业教师积极探索理学教育模式，“以学生为主体，教师为主导”进行教学改革与教学研究，取得了令人瞩目的成绩。我院公开出版教材23本（其中15本为近3年出版）。我专业采用多种模式的教学模式，独具特色的与企业的“校企联合培养计划”。</font><br>
    <br>
    <br>
    </body>
    </html>
```

feature.html 代码如下：

```
<%@ page pageEncoding="gb2312"%>
<html>
    <head>
        <title>feature.html</title>
        <meta http-equiv="keywords" content="keyword1,keyword2,keyword3">
        <meta http-equiv="description" content="this is my page">
        <meta http-equiv="content-type" content="text/html; charset=gb2312">
    </head>

    <body>
<font size="4" face="华文新魏">
    学院现有本科生739人，研究生106人。学院现有实验教学面积5000平米。在硬件条件上，经过天津市连续5个五年计划的投入、中央与地方共建项目以及学校实验室专项建设等多次成规模的投资，学院的基础实验室和专业实验室逐步完善。目前学院的教学科研仪器设备总值达到4000万元，为“强化实践育人”和高水平科学研究奠定了基础。
</font>
</body>
</html>
```

contact.html 代码如下：

```
<%@ page pageEncoding="gb2312"%>
<html>
```

```
<head>

<title>contact.html</title>

    <meta http-equiv="keywords" content="keyword1,keyword2,keyword3">
    <meta http-equiv="description" content="this is my page">
    <meta http-equiv="content-type" content="text/html; charset=gb2312">
</head>

<body>
<p>通讯地址：天津市津南区海河教园区职业技术学院</p>
<p> 物理与电子信息学院</p>
<p>邮政编码：300000 办公地点： 学校行政楼507</p>
<p>Email：zyjsxy@tjzy.edu.cn 电话：27778777</p>
<p>网址：http://tjzy.edu.cn/wl/</p>
<br>
  </body>
</html>
```

4. JSP 页面功能设计

（1）welcome.jsp 功能设计。在 chapter2 工程中，创建 welcome.jsp 页面，其功能是将 top.html 和 welcome.html 拼接成一个网页显示。代码如下：

```
<%@ page language="java" import="java.util.*" pageEncoding="gb2312"%>
<%
String path = request.getContextPath();
String basePath = request.getScheme()+"://"+request.getServerName()+":"+request.getServerPort()+
path+"/";
%>
<html>
  <head>
    <base href="<%=basePath%>">
   <title>物理与电子信息学院网站</title>
    <meta http-equiv="pragma" content="no-cache">
    <meta http-equiv="cache-control" content="no-cache">
    <meta http-equiv="expires" content="0">
    <meta http-equiv="keywords" content="keyword1,keyword2,keyword3">
    <meta http-equiv="description" content="This is my page">
  </head>
  <body>
   <!--以下内容来自文件top.html-->
   <%--inculde指令表示在编译时将文件插入到当前位置 --%>
   <%@ include file="top.html"%>
   <!--以下内容来自文件welcome.html-->
   <%@ include file="welcome.html" %>
  </body>
</html>
```

（2）feature.jsp 功能设计。在 chapter2 工程中，创建 feature.jsp 页面，其功能是将 top.html 和 feature.html 拼接成一个网页显示。代码如下：

```
<%@ page language="java" import="java.util.*" pageEncoding="gb2312"%>
<%
String path = request.getContextPath();
String basePath = request.getScheme()+"://"+request.getServerName()+":"+request.getServerPort()+
path+"/";
%>
<html>
  <head>
    <base href="<%=basePath%>">
   <title>物理与电子信息学院网站</title>
    <meta http-equiv="pragma" content="no-cache">
    <meta http-equiv="cache-control" content="no-cache">
    <meta http-equiv="expires" content="0">
    <meta http-equiv="keywords" content="keyword1,keyword2,keyword3">
    <meta http-equiv="description" content="This is my page">
  </head>
  <body>
   <!--以下内容来自文件top.html-->
   <%--inculde指令表示在编译时将文件插入到当前位置 --%>
   <%@ include file="top.html"%>
   <!--以下内容来自文件feature.html-->
   <%@ include file="feature.html" %>
  </body>
</html>
```

（3）contact.jsp 功能设计。在 chapter2 工程中，创建 contact.jsp 页面，其功能是将 top.html 和 contact.html 拼接成一个网页显示。代码如下：

```
<%@ page language="java" import="java.util.*" pageEncoding="gb2312"%>
<%
String path = request.getContextPath();
String basePath = request.getScheme()+"://"+request.getServerName()+":"+request.getServerPort()+
path+"/";
%>
<!DOCTYPE HTML PUBLIC "-//W3C//DTD HTML 4.01 Transitional//EN">
<html>
  <head>
    <base href="<%=basePath%>">
   <title>物理与电子信息学院网站</title>
    <meta http-equiv="pragma" content="no-cache">
    <meta http-equiv="cache-control" content="no-cache">
    <meta http-equiv="expires" content="0">
    <meta http-equiv="keywords" content="keyword1,keyword2,keyword3">
    <meta http-equiv="description" content="This is my page">
```

```
        </head>
    <body>
     <!--以下内容来自文件head.html-->
     <%--inculde指令表示在编译时将文件插入到当前位置 --%>
     <%@ include file="top.html"%>
     <!--以下内容来自文件welcome.html-->
     <%@ include file="contact.html" %>
    </body>
</html>
```

（4）index.jsp 功能设计。在 top.html 中，“学院首页”指向的链接为 index.jsp?choice=1，其中，index.jsp 表示要链接到的页面；链接地址后的？表示后面部分为所传递的参数；choice 表示传递的参数名；1 表示传递参数的参数值。如传递多个参数，则用&进行连接。相应的，“学院特色”指向的链接为 index.jsp?choice=2；“联络方式”指向的链接为 index.jsp?choice=3。index.jsp 通过 request.getParameter("choice")获取参数值，并将相应的页面嵌入到 index.jsp 的当前位置。代码如下：

```
<%@ page language="java" import="java.util.*" pageEncoding="gb2312"%>
<%
String path = request.getContextPath();
String basePath = request.getScheme()+"://"+request.getServerName()+":"+request.getServerPort()+
path+"/";
%>
<!DOCTYPE HTML PUBLIC "-//W3C//DTD HTML 4.01 Transitional//EN">
<html>
  <head>
    <base href="<%=basePath%>">
   <title>物理与电子信息学院网站</title>
    <meta http-equiv="pragma" content="no-cache">
    <meta http-equiv="cache-control" content="no-cache">
    <meta http-equiv="expires" content="0">
    <meta http-equiv="keywords" content="keyword1,keyword2,keyword3">
    <meta http-equiv="description" content="This is my page">
      </head>
  <body>
   <%@ include file="top.htm"%>
   <%
     //从超级链接获取参数值
     String s=request.getParameter("choice");
     //如果没有获取到，则默认为“1”
     if (s==null)
     s="1";
     int choice=Integer.parseInt(s);
     switch(choice){
      case 1:     %>
 <!--专业频道，引入welcome.html  -->
```

```
                <%@ include file="welcome.html"%>
            <%break;
            case 2:%>
<!--特色频道，引入feature.html  -->
                <%@ include file="feature.html"%>
            <%break;
            case 3:%>
<!--联络频道，引入contact.html  -->
                <%@ include file="contact.html"%>
                <%break;
                }
        %>
    </body>
</html>
```

本章小结

任何语言都有自己的语法，JSP 也不例外。作为以 Java 为基础的一种动态网页编程技术，JSP 页面除了包括 HTML 静态元素和包含 Java 代码段的脚本标记外，还拥有功能丰富的指令标记和动作标记等。了解页面的构成，掌握语法规范，是应用 JSP 动态 Web 技术的重要基础。

课后练习

1. 简述 JSP 页面的构成。
2. 简述 JSP 静态注释和动态注释的区别及用法。
3. 简述指令标记 include 和动作标记 include 的区别。

PART03

项目三 JSP内置对象

JSP 内置对象是不需要声明和创建就可以在 JSP 页面脚本中使用的成员变量。通过这些内置对象，可以实现响应客户端的请求、向客户端发送数据等功能。本章将详细介绍内置对象的使用方法。

课堂学习目标

- 掌握 JSP 内置对象的语法规范
- 精通 JSP 内置对象的应用
- 熟练使用 JSP 开发环境

素养拓展

- 建立自己解决问题的“方法库”

素养拓展

任务一　JSP 内置对象简介

任务要求

本任务要求理解 JSP 中 9 个内置对象的主要功能，了解它们之间的相互联系。

任务实现

（一）JSP 内置对象概述

JSP 提供了 9 个内置对象，这些对象在 JSP 环境下不需要预先声明和创建就能直接使用。这 9 个内置对象分别是 request、response、application、session、out、config、pageContext、page、exception，其功能如表 3-1 所示。

表 3-1　JSP 页面的 9 个内置对象

对象名称	衍生类	功能简述
request	javax.servlet.ServletRequest.HttpServlet Request	取得客户端与系统的信息
response	javax.servlet.ServletRequest.HttpServlet Response	响应客户端信息
application	javax.servlet.ServletContext	记录与处理上线者共享的数据
session	javax.servlet. http.HttpSession	记录与处理上线者的个别数据
out	javax.servlet.jsp.JspWriter	控制数据输出的操作
config	javax.servlet.ServletConfig	取得 JSP 编译后的 Servlet 信息
pageContext	javax.servlet.jsp.PageContext	存取和处理系统运行时的各种信息
page	java.lang.Object	代表目前的这个 JSP 网页对象
exception	java.lang.Throwable	异常处理机制

（二）内置对象之间的联系

1. request 内置对象与 response 内置对象

JSP 页面之所以具备与用户交互的功能，关键在于 request 内置对象与 response 内置对象，request 内置对象让服务器取得用户在网页表单中输入的数据内容，response 内置对象则提供服务器端程序响应客户端信息所需的功能。

2. application 内置对象与 session 内置对象

application 与 session 这两个内置对象主要用于记录和处理 JSP 页面之间的共享数据。由于因特网本身是一种无联机状态的应用程序，当一份网页文件从网站服务器传至客户端的浏览器之后，客户端与服务器之间就没有任何联机状态存在，这个先天缺陷让网页无法存储应用程序运行期间所需的共享数据，application 内置对象与 session 内置对象就是用来解决这样的问题的。

3. out 内置对象

JSP 页面是一种动态的网页，与 HTML 这一类静态文件的最大不同在于，同一网页经过程序运算得以根据各种条件及情况进行不同呈现。out 内置对象在这一方面提供相关支持，服务器端利用 out 内置对象将要输出的内容在传送至网页的时候动态地写入客户端。

4. config、pageContext 以及 page 内置对象

这三个内置对象用于存取 JSP 页面运行阶段的各种信息内容，其中 config 内置对象包含 JSP 页面被编译成为 Servlet 之后的相关信息；pageContext 内置对象则是提供系统运行期间的各种信息内容的存取操作功能；page 内置对象代表目前正在运行的 JSP 网页对象。

5. exception 内置对象

exception 内置对象为 JSP 提供用于处理程序运行错误的异常对象，此对象搭配功能强大的异常处理机制，运用于 JSP 网页的程序除错和异常处理上。

（三）内置对象的生命周期

session、application、pageContext 和 request 内置对象实现数据在网页间的传递，但其作用域各不相同，JSP 提供了四种属性的保存范围：

page 设置的属性只在当前页面有效；

request 设置的属性在一次请求范围内有效；

session 设置的属性的有效期在客户浏览器与服务器一次会话范围内，如果服务器断开连接，该属性就失效了；

application 设置的属性在服务器开启时执行服务，直到服务器关闭为止。

各内置对象对应的作用域见表 3-2。

表 3-2　内置对象的作用域

内置对象	作用域
request 对象	request
response 对象	page
pageContext 对象	page
session 对象	session
application 对象	application
out 对象	page
config 对象	page
page 对象	page
exception 对象	page

任务二　request 内置对象

任务要求

本任务要求认知 request 内置对象的常用方法，并熟练掌握 request 内置对象的应用。

任务实现

（一）request 内置对象的常用方法

request 内置对象主要用于接收客户端通过 HTTP 协议连接传输到服务器端的数据，它通常是 HttpServletRequest 的子类，其作用域就是一次 request 请求。request 内置对象的常用方法如表 3-3 所示。

表 3-3　request 内置对象的常用方法

方法	说明
getAttribute(String name)	返回 name 所指定的属性值
setAttribute(String name, Object obj)	设定 name 所指定的属性值为 obj
removeAttribute(String name)	删除 name 所指定的属性
getAttributeNames()	返回 request 对象所有的属性名称集合
getParameter(String name)	从客户端获取 name 所指定的参数值
getParameter Names()	从客户端获取所有参数名称
getParameterValues(String name)	从客户端获取 name 所指定参数的所有值
setCharacterEncoding(String encoding)	设定请求正文中所使用的字符编码（只支持 post 提交的数据）

（二）request 内置对象应用实例

【例 3-1】 利用 request 内置对象获取信息并显示的实例。

本例包括两个文件：requestLogin.jsp 和 requestShowInfo.jsp。在 requestLogin.jsp 页面中输入用户名和密码，在 requestShowInfo.jsp 页面中利用 request 内置对象的方法将输入的用户名、密码等信息显示出来。

requestLogin.jsp 的代码如下所示：

```
<%@ page contentType="text/html;charset=GB2312" language="java"
    pageEncoding="GB2312" %>
<html>
<head>
    <meta http-equiv="Content-Type" content="text/html; charset=GB2312">
    <title>使用request内置对象--requestLogin.jsp</title>
</head>
<body>
<body bgcolor="#ffc7c7">
<form name="form1" method="post" action="requestShowInfo.jsp">
   <p align="center">用户名:<input type="text" name="username"></p>
   <p align="center">密　码: <input type="password" name="password"></p>
   <p align="center">
      <input type="submit" name="Submit" value="OK">
      <input type="reset" name="Reset" id="Reset" value="Cancel">
```

```
        </p>
    </form>
    </body>
    </body>
    </html>
```

requestShowInfo.jsp 的代码如下所示：

```
<%@ page contentType="text/html;charset=GB2312" language="java"
    pageEncoding="GB2312" import="java.util.*" %>
<html>
<head>
    <meta http-equiv="Content-Type" content="text/html; charset=GB2312">
    <title>使用request内置对象--requestShowInfo.jsp</title>
</head>
<body bgcolor="#ccffcc">
<h1>您刚才输入的内容是：<br></h1>
<%
    Enumeration enu = request.getParameterNames(); //使用Enumeration时，需要在page标签里
import="java.util.*"
    while(enu.hasMoreElements()) {
        String parameterName = (String)enu.nextElement();
        String parameterValue = request.getParameter(parameterName);
        out.print("参数名称："+parameterName+"<br>");
        out.print("参数内容："+parameterValue+"<br>");
    }
%>
</body>
</html>
```

requestLogin.jsp 的运行效果如图 3-1 所示。单击 OK 按钮后出现如图 3-2 所示的结果。

图 3-1

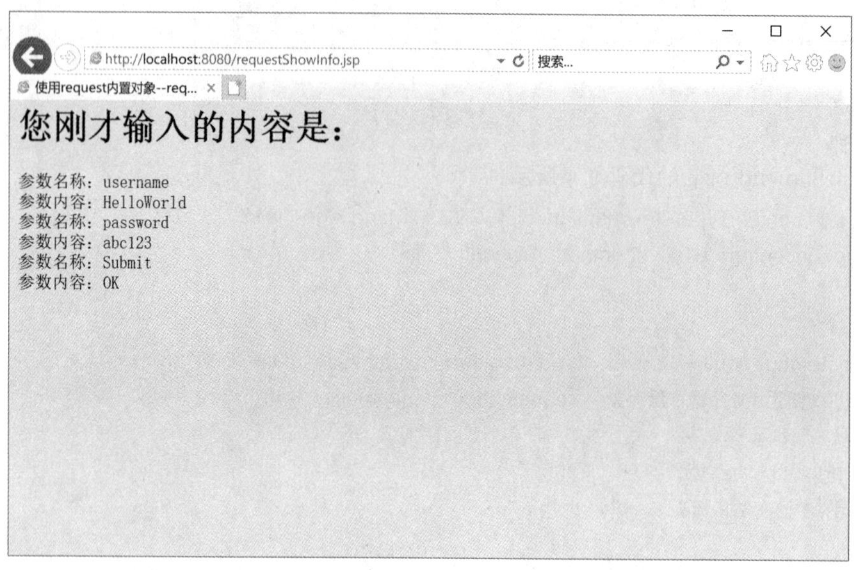

图 3-2

任务三　response 内置对象

任务要求

本任务要求认知 response 内置对象的常用方法，并熟练掌握 response 内置对象的应用。

任务实现

（一）response 内置对象的常用方法

response 内置对象用于将服务器端数据发送给客户端以响应客户端的请求。response 内置对象的常用方法如表 3-4 所示。

表 3-4　response 内置对象的常用方法

方法	说明
setContentType(String type)	动态响应 contentType 属性
setHeader(String name，String value)	设置 HTTP 应答报文的首部字段和值及自动更新
sendRedirect(String redirectURL)	将客户端重定向到指定 URL
setStatus(int n)	设置 HTTP 返回的状态值
addCookie(Cookie cookie)	添加一个 Cookie 对象

（二）rseponse 内置对象应用实例

【例 3-2】 利用 response 内置对象控制刷新频率的实例。

在 responseRefresh.jsp 页面中设置页面刷新的频率，并在页面中实时显示当时的时间。

responseRefresh.jsp 的代码如下所示：

```
<%@ page contentType="text/html;charset=GB2312" language="java" %>
<html>
<head>
    <meta http-equiv="Content-Type" content="text/html; charset=GB2312">
    <title>使用response内置对象--responseRefresh.jsp</title>
</head>
<body>
    <h3>现在的时间是：</h3>
    <hr/>
    <%=new java.util.Date()%>
    <%
        response.setHeader("refresh", "1");//对属性refresh赋值，页面每一秒刷新一次
    %>
    <hr/>
</body>
</html>
```

responseRefresh.jsp 的运行效果如图 3-3 所示。

图 3-3

任务四　out 内置对象

任务要求

本任务要求认知 out 内置对象的常用方法，并熟练掌握 out 内置对象的应用。

任务实现

（一）out 内置对象的常用方法

out 内置对象是 JspWriter 类的实例，是向客户端输出内容常用的对象。out 内置对象的常用方法如表 3-5 所示。

表 3-5　out 内置对象的常用方法

方法	说明
clear()	清除缓冲区中的数据，若缓冲区已经是空的，则会产生 IOException 异常
clearBuffer()	清除缓冲区中的数据，若缓冲区已经是空的，并不会产生 IOException 异常
flush()	直接将暂存于缓冲区中的数据清空并输出到网页
getBufferSize()	返回缓冲区的大小
getRemaining()	返回缓冲区中剩余空间的大小
isAutoFlush()	返回布尔值，表示是否自动输出缓冲区中的数据
newline()	输出换行
print(datatype data)	输出数据类型为 datatype 的数据 data
println(datatype data)	输出数据类型为 datatype 的数据 data，并且自动换行（下一个输出语句在下一行开始输出）

（二）out 内置对象应用实例

【例 3-3】 利用 out 内置对象进行输出的实例。

本例在 out.jsp 中利用 out 内置对象的 println(datatype data)方法将指定内容输出，并利用 out 内置对象的 getBuffersize()和 getRemaining()方法分别获取缓冲区及其剩余空间的大小。

out.jsp 的代码如下所示：

```
<%@ page contentType="text/html;charset=GB2312" language="java" %>
<html>
<head>
    <title>使用out内置对象--out.jsp</title>
</head>
<body>

<h2>out内置对象</h2>
<%
    out.println("学习使用out内置对象:<br>");
    int Buffer = out.getBufferSize();
    int Available = out.getRemaining();
%>
输出缓冲区的大小 ：<%= Buffer %><br>
缓冲区剩余空间的大小 ：<%= Available %><br>
```

```
</body>
</html>
```

out.jsp 页面的运行效果如图 3-4 所示。

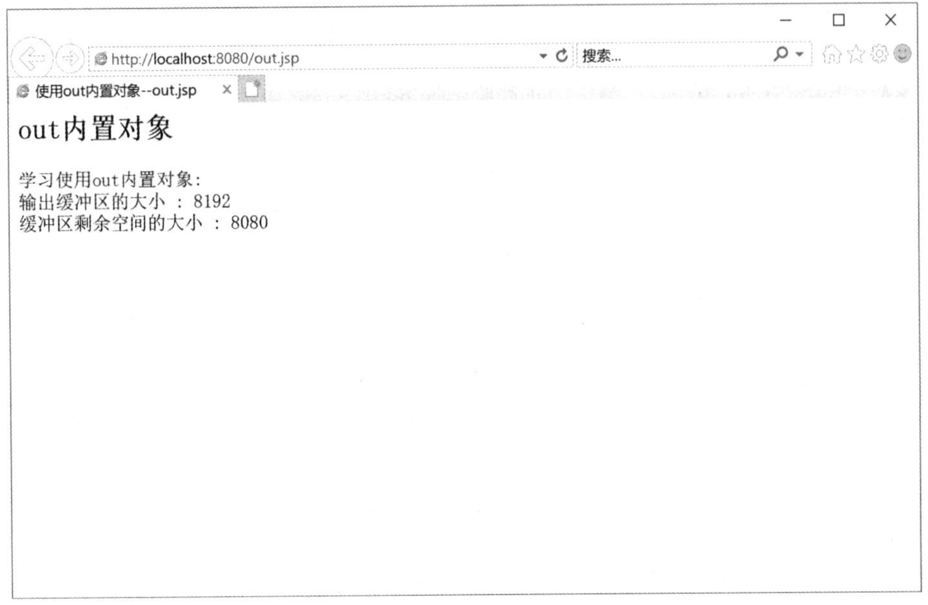

图 3-4

任务五　session 内置对象

任务要求

本任务要求认知 session 内置对象的常用方法，并熟练掌握 session 内置对象的应用。

任务实现

（一）session 的概念

session（会话）一词其本来的含义是指有始有终的一系列动作/消息，比如打电话时从拿起电话拨号到挂断电话这一系列过程可以称之为一个 session。在网络应用中，session 对象存储特定用户会话的属性及配置信息。这样，当用户在应用程序的 Web 页之间跳转时，存储在 session 中的变量将不会丢失，而是在整个用户会话中一直存在。当用户请求来自应用程序的 Web 页面时，如果该用户还没有会话，则 Web 服务器将自动创建一个 session 对象。当会话过期或被放弃后，服务器将终止该会话。

（二）session 内置对象的常用方法

session 内置对象的常用方法如表 3-6 所示。

表 3-6　session 内置对象的常用方法

方法	说明
setAttribute(String name，Object obj)	在 session 中设定 name 所指定的属性值为 obj
getAttribute(String name)	返回 session 中 name 所指定的属性值
getAttributeNames()	返回 session 中所有变量的名称
removeAttribute(String name)	删除 session 中 name 所指定的属性
invalidate()	销毁与用户关联的 session
getCreationTime()	返回 session 建立的时间，返回值为从格林尼治时间 1970 年 1 月 1 日开始算到 session 建立时的时间（单位毫秒）
getLastAccessedTime()	返回客户端对服务器端提出请求至处理 session 中数据的最后时间，若为新建的 session，则返回-1
getMaxInactiveInterval()	返回客户端未对 session 开始停滞到自动消失之间所间隔的时间，返回值单位为秒
isNew()	返回布尔值，表示是否为新建的 session。新建是指程序调用 session 对象在服务器端建立 session，而尚未将此 session 的信息记录到客户端的 Cookie 中
setMaxInactiveInterval(int interval)	设置客户端未对 session 提出请求而 session 开始停滞到自动消失之间所间隔的时间为 interval，以秒为单位

（三）session 内置对象应用实例

【例 3-4】 利用 session 内置对象统计访问站点人数的实例。

本例在 sessionCount.jsp 中利用 session 内置对象中的 isNew()方法判断当前是否为一个新创的 session，如果是则访问数加 1，否则访问数不变。

seesionCount.jsp 的代码如下所示：

```
<%@ page contentType="text/html;charset=GB2312" language="java" %>
<html>
<head>
    <title>session内置对象计数器--sessionCount.jsp</title>
    <style type="text/css">
        h1{color:red; text-align: center; size:7px}
        p{color:green; text-align: center; size: 5px}
    </style>
</head>
<body>
<%!int Num = 0; %>
<%
    if(session.isNew()) {
      Num += 1;
      session.setAttribute("Num", Num);//将Num变量值存入session
    }
%>
```

```
<h1>session计数器</h1>
<br>
<p>
    您是第
    <%=session.getAttribute("Num") %>
    个访问本网站的用户
</p>
</body>
</html>
```

第一次运行 sessionCount.jsp 时，效果如图 3-5 所示。在浏览器中刷新页面时，因为没有创建新的会话，所以访问人数并不会有变化；当关闭浏览器（即关闭客户端与服务器当前对话）后，再次打开该页面，访问人数才会增加，如图 3-6 所示。

图 3-5

图 3-6

任务六　application 内置对象

任务要求

本任务要求认知 application 内置对象的常用方法，并熟练掌握 application 内置对象的应用。

任务实现

（一）application 内置对象的常用方法

前一个任务中介绍了在 JSP 页面中使用 session 来存储每个用户的私有信息，但是有时候服务器需要管理面向整个应用的参数，使得每个用户都能获得相同的参数值，这时就需要用到 application 内置对象。application 对服务器而言，可以视为一个所有联机用户共享的数据存取区，application 中的变量数据在程序设置其值时被初始化，而当网页服务器被关闭，或者超过预设时间而未有任何用户联机时将自动消失。对每一个联机浏览网页的用户来说，application 对象用于存储其共享数据，无论是网站中任何一个网页，用户存取的数据内容均相同，可以将其视为传统应用程序中的全局共享数据。

application 内置对象的常用方法如表 3-7 所示。

表 3-7　application 内置对象的常用方法

方法	说明
setAttribute(String name，Object obj)	application 中设定 name 所指定的属性值为 obj
getAttribute(String name)	返回 application 中 name 所指定的属性值
getAttributeNames()	返回 application 中所有变量的名称
removeAttribute(String name)	删除 application 中 name 所指定的属性
getMajorVersion()	返回服务器解释引擎所支持的最新 Servlet API 版本
getMinorVertion()	返回服务器解释引擎所支持的最低 Servlet API 版本
getMimeType(string file)	返回文件 file 的文件格式与编码方式
getRealPath(String path)	返回虚拟路径 path 的真实路径
getServerInfo()	返回服务器解释引擎的信息

（二）application 内置对象应用实例

【例 3-5】 利用 application 内置对象实现共享留言板的实例。

本例在 inputMessage.jsp 中通过表单呈现出留言板，并搜集用户填写的内容。在 checkMessage.jsp 中，接收表单中传递过来的信息并加以修饰，通过 application 内置对象中的 setAttribute(String name) 方法将用户填写的信息存入相应的全局变量。在 showMessage.jsp 中，通过 application 内置对象中的 getAttribute(String name)获取全局变量的值，并以适当的形式显示出来。

inputMessage.jsp 的代码如下：

```
<%@ page language="java" import="java.text.*,java.util.*"
```

```
            contentType="text/html; charset=UTF-8" pageEncoding="UTF-8"%>
<html>
<head>
    <meta http-equiv="Content-Type" content="text/html; charset=UTF-8">
    <title>使用application内置对象--inputMessage.jsp</title>
    <style>
        #form2 input {
            color: green;
            font-weight: bold;
        }
    </style>
</head>
<body bgcolor="#abcdef">

<form action="checkMessage.jsp" method="post">
    请输入姓名： <input type="text" name="name" /><br>
    请输入标题： <input type="text" name="title" /><br>
    请输入内容：
    <textarea cols="40" rows="10" name="message"></textarea>
    <br> <br> <br>
    <input type="submit" value="留言" />
</form>
<br>
<form id="form2" action="showMessage.jsp" method="post">
    <input type="submit" value="查看留言板" />
</form>

</body>
</html>
```

checkMessage.jsp 的代码如下：

```
<%@ page language="java" import="java.text.*,java.util.*"
            contentType="text/html; charset=UTF-8" pageEncoding="UTF-8"%>
<html>
<head>
    <meta http-equiv="Content-Type" content="text/html; charset=UTF-8">
    <title>使用application内置对象--checkMessage.jsp</title>
</head>
<body bgcolor="#abcdef">
<%!Vector<String> v = new Vector<String>();
    int i = 0;%>
<%
    String datetime = new SimpleDateFormat("yyyy-MM-dd hh:mm:ss").format(Calendar.getInstance().
getTime()); //获取系统时间
%>
<%
```

```
        request.setCharacterEncoding("utf-8");
        String name = request.getParameter("name");
        String title = request.getParameter("title");
        String message = request.getParameter("message");
    %>
    <%
        if (name == null || "".equals(name.trim())) {
            //trim()主要解决里面只有空格的问题
            name = " 网友" + (int) (Math.random() * 100000 + 10000);
        }
        if (title == null || "".equals(title.trim())) {
            title = " 无";
        }
        if (message == null || "".equals(message.trim())) {
            message = " 无";
        }
    %>
    <%
        i++;
        String str = "第" + "<span class=span0>" + i + "</span> " + "楼  "
                + ".<span class=span1>留言人: </span>" + name + ".<span class=span2>标题: </span>" +
title
                + ".<span class=span3>内容: </span><br>       " + message
                + ".<span class=span4>时间: </span>   " + datetime + ".<hr>";

        v.add(str);
        application.setAttribute("message", v);
    %>
    留言成功.
    <a href="inputMessage.jsp">返回留言板</a>
    </body>
    </html>
```

showMessage.jsp 的代码如下：

```
<%@page import="com.sun.org.apache.xml.internal.serializer.utils.StringToIntTable"%>
<%@ page language="java" import="java.util.*"
        contentType="text/html; charset=UTF-8" pageEncoding="UTF-8"%>
<html>
<head>
    <meta http-equiv="Content-Type" content="text/html; charset=UTF-8">
    <title>使用application内置对象--showMessage.jsp</title>
    <style>
        body {
            background: RGBA(38, 38, 38, 1);
        }
        div {
```

```
            width: 800px; //
        border: 1px solid RGBA(100, 90, 87, 1);
            color: white;
        }
        span {
            font-size: 20px;
            font-weight: bold;
        }
        .span0 {
            color: red;
            font-size: 25px;
        }
        .span1 {
            color: green;
        }
        .span2 {
            color: orange;
        }
        .span3 {
            color: green;
        }
        .span4 {
            color: red;
        }
    </style>
</head>
<body>
<div>
    <%
        Object o = application.getAttribute("message");
        if (o == null) {
          out.print("暂时还没有留言呢");
        } else {
          Vector<String> v = (Vector<String>) o;
          for (int i = v.size() - 1; i >= 0; i--) {
              // 注意必须用/.  String[] st1 = v.get(i).split("/.");
              //                for (int j = 0; j < st1.length; j++) {
              //                     out.print(st1[j] + "<br>");
              //                }
              //                out.print("<br>");

              StringTokenizer st = new StringTokenizer(v.get(i), ".");
              while (st.hasMoreElements()) {
                  out.print(st.nextToken() + "<br>");
              }
```

```
            }
        }
    %>
</div>
</body>
</html>
```

运行 inputMessage.jsp 的效果如图 3-7 所示，用户填写相应的信息后单击“留言”按钮，则会将信息传送至 checkMessage.jsp，运行效果如图 3-8 所示。单击“返回留言板”链接，就会通过超级链接返回到留言界面（inputMessage.jsp 页面）。在 inputMessage.jsp 页面中单击“查看留言板”按钮，则可以查看所有用户的历史留言，如图 3-9 所示。

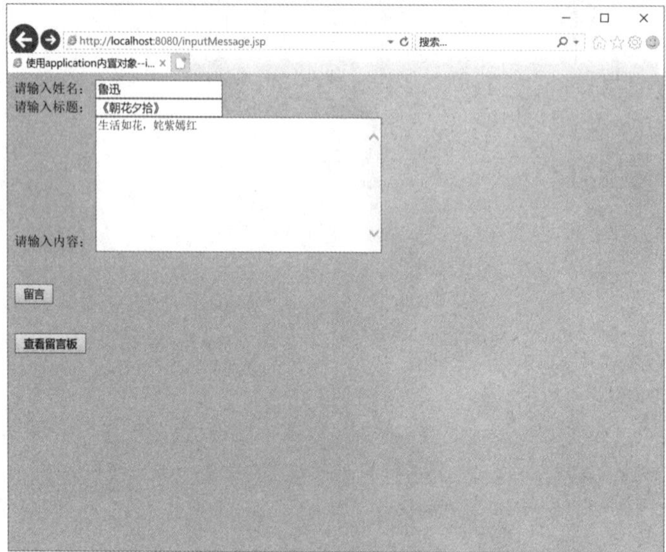

图 3-7

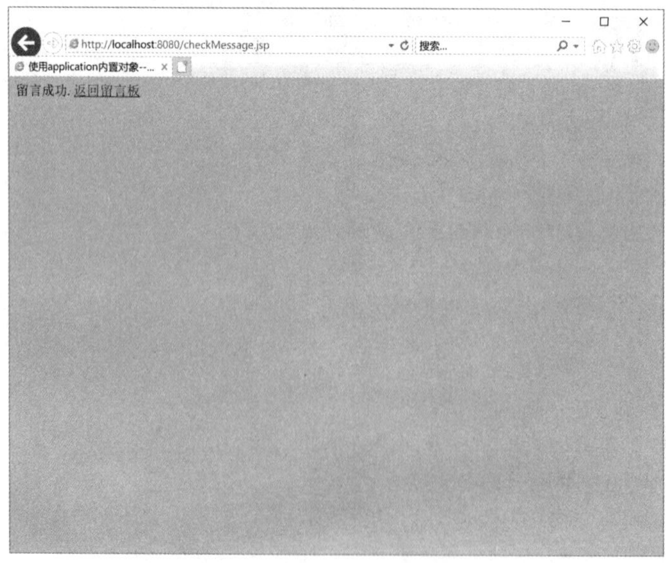

图 3-8

图 3-9

任务七　拓展实训

任务要求

完成一个网上调查网站，进一步掌握 JSP 的内置对象。

任务实现

（一）问卷调查网页的运行效果

人们在日常生活当中经常会参与一些问卷调查。本任务是完成一个类似网上问卷调查的网站，如图 3-10～图 3-13 所示。

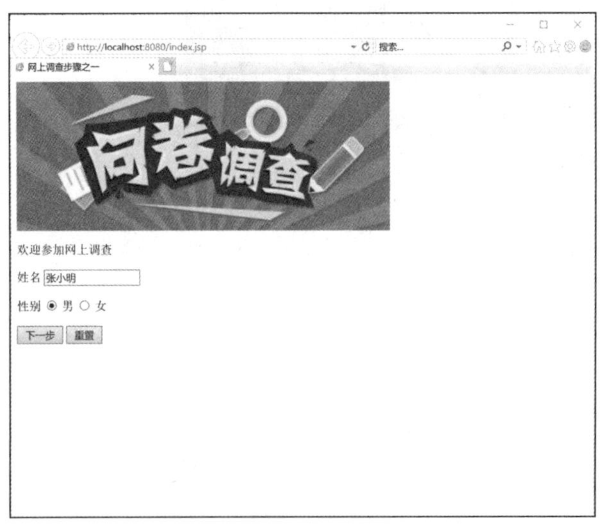

图 3-10

在图 3-10 中，首先需要受访者填写姓名并选择性别。在图 3-11 中，受访者选择自己的业余爱好，单击“下一步”按钮时，页面能够显示出用户选择的信息，同时将当前调查的统计结果显示出来，如图 3-12、图 3-13 所示。

图 3-11

图 3-12

图 3-13

（二）功能设计

1. 创建工程

启动 IntelliJ IDEA，创建 Web 应用并命名为 chapter3。展开工程，在 web 文件夹下创建 images 子文件夹，并将 top.jpg 导入到 images 文件夹中。在 web 文件夹下创建 JSP 类型文件 index.jsp、hobby.jsp、result.jsp 和 HTML 类型文件 top.html。

2. top.html 功能设计

如图 3-10～图 3-13 所示，该问卷调查网站每个页面的题头部分都有相同的 Logo 图片，将这一部分单独写成一个页面 top.html。将 Logo 图片 top.jpg 插入到 top.html 中，并通过 include 指令标记将 top.html 插入到每一个网页中。

3. index.jsp 功能设计

主页面 index.jsp 的运行效果如图 3-10 所示，其中，首先通过 page 指令设定页面的相关属性，并通过 include 指令标记将 top.html 插入到页面中，再建立表单，将“姓名”文本框命名为 name，将“性别”单选按钮命名为 sex，取值分别为 male 和 female，默认值为“男”，处理表单信息的页面为 hobby.jsp。代码如下：

```
<%@ page contentType="text/html;charset=UTF-8" language="java" %>
<html>
<head>
  <title>网上调查步骤之一</title>
  <meta http-equiv="description" content="questionaire">
  <meta http-equiv="content-type" content="text/html; charset=UTF-8">
</head>
```

```
<body>
<%@include file="top.html"%>
<form name="form1" method="post" action="hobby.jsp">
  <p>欢迎参加网上调查</p>
  <p>姓名
    <input name="name" type="text" id="name" size="16">
  </p>
  <p>性别
    <input name="sex" type="radio" value="male" checked>
    男
    <input name="sex" type="radio" value="female">
    女</p>
  <p>
    <input type="submit" name="Submit" value="下一步">
    <input type="reset" name="Submit2" value="重置">
  </p>
</form>
</body>
</html>
```

4. hobby.jsp 功能设计

在 hobby.jsp 中，获取 index.jsp 页面传递过来的信息，并保存在 session 中，并通过表达式在“，你好，请继续完成调查”的前面将受访者的信息显示出来，如图 3-11 所示。代码如下：

```
<%@ page language="java" import="java.util.*" pageEncoding="UTF-8"%>
<%
    String path = request.getContextPath();
    String basePath = request.getScheme()+"://"+request.getServerName()+":"+
request.getServerPort()+path+"/";
%>
<html>
<head>
    <base href="<%=basePath%>">
    <title>网上调查步骤之二</title>
</head>
<body>
<%
    String name=request.getParameter("name");
    name=new String(name.getBytes("ISO-8859-1"),"UTF-8");
    String sex=request.getParameter("sex");
    if(sex.equals("male"))
      sex="先生";
    else sex="女士";
    session.putValue("namesex",name+sex);
%>
<%@include file="top.html"%>
```

```
<p>
//显示受访者的姓名和性别
    <%=name%>
    <%=sex%>
    ，你好，请继续完成调查 </p>
<p>你感兴趣的业余爱好有：</p>
<form name="form1" method="post" action="result.jsp">
    <p>
      <input name="hobby0" type="checkbox" id="film" value="影视欣赏">
      影视欣赏</p>
    <p>
      <input name="hobby1" type="checkbox" id="book" value="阅读书籍">
      阅读书籍</p>
    <p>
      <input name="hobby2" type="checkbox" id="sports" value="体育运动">
      体育运动</p>
    <p>
      <input name="hobby3" type="checkbox" id="travel" value="户外旅游">
      户外旅游</p>
    <p>
      <input name="hobby4" type="checkbox" id="music" value="流行音乐">
      流行音乐</p>
    <p>
      <input type="submit" name="Submit" value="下一步">
      <input type="reset" name="Reset" value="重置">
    </p>
</form>
</body>
</html>
```

5. result.jsp 功能设计

在 result.jsp 中，获取 hobby.jsp 页面传递过来的信息，由于需要对所有受访者的选择结果进行统计，因此需要将信息保存到 application 中，以实现不同用户之间统计数据的共享。将存在 session 中的用户姓名及性别显示在“，你所感兴趣的项目如下：”之前，并利用循环将数组中存储的受访者选择显示出来。最后，将存储在 application 中的调查历史结果显示出来。代码如下：

```
<%@ page language="java" import="java.util.*" pageEncoding="UTF-8"%>
<%
    String path = request.getContextPath();
    String basePath = request.getScheme()+"://"+request.getServerName()+":"+request.
getServerPort()+path+"/";
%>
<html>
<head>
    <base href="<%=basePath%>">
```

```
    <title>网上调查步骤之三</title>
</head>
<body>
<%@include file="top.html"%>
<%
    String namesex=(String) session.getValue("namesex");
    String hobby[]=new String[5];
    for (int i=0;i<hobby.length;i++){
//new数组中存储了受访者的选择
        String param="hobby"+i;
        String getparam=request.getParameter(param);
        if(getparam!=null){
            hobby[i]=new String(getparam.getBytes("ISO-8859-1"),"UTF-8");
            synchronized (application){
                Integer count=(Integer) application.getAttribute(param);
                if(count==null)
                    count=new Integer(0);
                count=new Integer(count.intValue()+1);
                application.setAttribute(param,count);
            }
        }
    }

%>
<p>
    <%=(String) session.getValue("namesex") %>
    ，你所感兴趣的业余爱好如下：</p>
<%
    for (int i=0;i<hobby.length;i++)
        if(hobby[i]!=null){
            out.print("<p>");
            out.println(hobby[i]);
            out.println("</p>");
        }
    String hobbyname[]={"电影欣赏","阅读书籍","体育运动","户外旅游","流行音乐"};
    for (int i=0;i<hobby.length;i++){
        Integer count=(Integer) application.getAttribute("hobby"+i);
        if(count==null)
            count=new Integer(0);
        out.println("<p>选择"+hobbyname[i]+"的人次为"+count+"</p>");
    }
%>
</body>
</html>
```

本章小结

JSP 引擎在调用 JSP 对应的 jspServlet 时，会传递或创建 9 个与 Web 开发相关的对象供 jspServlet 使用。JSP 的设计者为了便于开发人员在编写 JSP 页面时获得和这些 Web 对象的应用，特意定义了 9 个相应的变量，开发人员在 JSP 页面中通过这些变量就可以快速获得这九大对象的引用，这就是所谓的内置对象。运用好 JSP 内置对象，不仅可以提升编程的效率，同时可以有效实现用户与网页的动态交互。

课后练习

1. 简述 JSP 九大内置对象的基本功能。
2. 试比较 seesion 内置对象、page 内置对象、request 内置对象和 application 内置对象的作用范围。
3. 将例 3-4 中的“session”换成“application”，运行效果会有何不同？为什么会有这样的差异？
4. 怎样使用 request 内置对象、session 内置对象、application 内置对象进行参数存取？

PART04

项目四

JDBC技术的应用

Web 开发不可避免地要使用数据库来存储和管理数据。为了在 Java 语言中提供对数据库访问的支持，Sun 公司于 1996 年提供了一套访问数据库的标准 Java 类库，即 JDBC（Java Data Base Connectivity，Java 数据库连接）。本章基于 JDBC 技术，讲解对数据库的增删改查，然后讲解数据库操作过程中应该注意的一些问题。

课堂学习目标

- 了解 JDBC 技术
- 掌握使用 JDBC 访问数据库的基本步骤
- 了解常见的数据库连接方式

素养拓展

- 守序与和谐

素养拓展

任务一　JDBC 技术概述

任务要求

本任务要求了解 JDBC 的基本知识。

任务实现

在前面的章节中知道，JSP 中可以写 Java 代码，很明显可以通过 Java 代码来访问数据库。在 Java 技术系列中，访问数据库的技术叫作 JDBC，它提供了一系列的 API，用 Java 语言编写的代码连接数据库，对数据库的数据进行添加、删除、修改和查询。应用程序使用 JDBC 访问数据库的方式如图 4-1 所示。

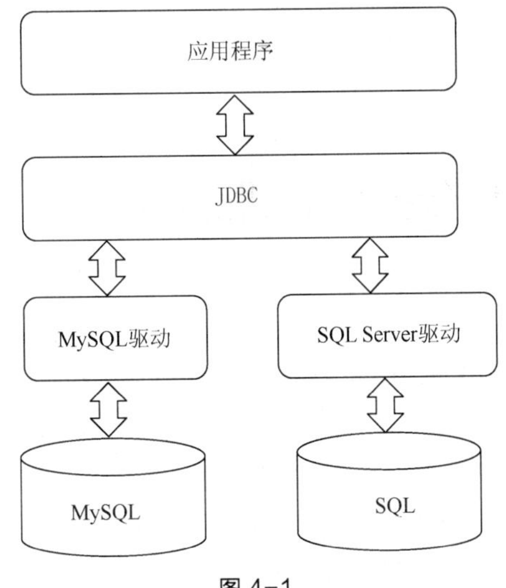

图 4-1

不过，这里有一个问题。由于 JSP 不知道具体连接的是哪一种数据库，而各种数据库产品由于厂商不一样，连接的方式也不一样。因此，为了使应用程序与数据库真正建立连接，JDBC 不仅需要提供访问数据库的 API，还需要封装与各种数据库服务器通信的细节。

数据库开发人员使用 JDBC API 编写一个程序后，就可以很方便地将 SQL 语句传送给几乎任何一种数据库，如 Sybase、Oracle 或 Microsoft SQL Server 等。用 JDBC 写的程序能够自动地将 SQL 语句传送给相应的数据库管理系统。Java 和 JDBC 的结合可以让数据库开发人员在开发数据库应用时真正实现“只写一次，随处运行”。

JDBC API 既支持数据库访问的两层模型，同时也支持三层模型。

在两层模型中，Java Applet 或应用程序将直接与数据库进行对话。在这种情况下，需要一个 JDBC 驱动程序来与所访问的特定数据库管理系统进行通信。用户的 SQL 语句被送往数据库中，而处理的结果将被送回给用户。存放数据的数据库可以位于另一台物理计算机上，用户通过网络连接到数据库服务器，这就是典型的客户/服务器模型，其中用户的计算机为客户机，提供数据库的计算机为服务器。网络可以是公司内部的 Intranet， 也可以是 Internet。两层模型如图 4-2 所示。

在三层模型中，命令先是被发送到服务的“中间层”，然后由它将 SQL 语句发送给数据库，数据库对 SQL 语句进行处理并将结果送回到中间层，中间层再将结果送回给用户。三层模型如图 4-3 所示。

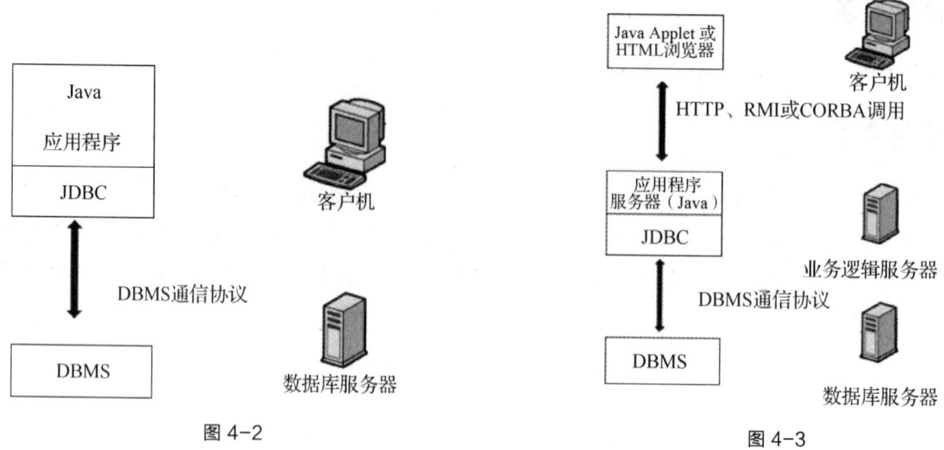

图 4-2

图 4-3

任务二　MySQL 数据库管理系统

任务要求

本任务要求掌握安装 MySQL 数据库的过程，能够正确操作数据库。

任务实现

MySQL 是当今最流行的关系型数据库管理系统，在 Web 应用方面 MySQL 是最好的 RDBMS（Relational Database Management System，关系数据库管理系统）应用软件之一，目前属于 Oracle 公司。关系型数据库将数据保存在不同的表中，而不是将所有数据放在一个大仓库内，这样就提高了速度并提高了灵活性。MySQL 是一个真正的多用户、多线程 SQL 数据库服务器。SQL（结构化查询语言）是世界上最流行的和标准化的数据库语言，它使得存储、更新和存取信息更加容易。

与其他的大型数据库如 Oracle、DB2、SQL Server 等相比，MySQL 自有它的不足之处，但是这丝毫也没有减少它受欢迎的程度。对于一般的个人使用者和中小型企业来说，MySQL 提供的功能已经绰绰有余，而且由于 MySQL 是开放源码软件，因此可以大大降低开发的总体成本。

（一）安装 MySQL 数据库

MySQL 是开源项目，很多网站都提供免费下载。可以使用任何搜索引擎搜索关键字“MySQL 下载”，来获得有关的下载地址。这里选择的地址是 MySQL 的官方网站 https://dev.mysql.com/downloads/，该网站免费提供 MySQL 最新版本的下载以及相关技术文章。

1. MySql 的安装步骤

（1）将压缩包解压到某个目录，比如 C:\mysql。

（2）双击 C:\mysql 下的 setup.exe 开始安装。

（3）运行安装包，勾选同意，在该界面上单击 Next 按钮，如图 4-4 所示。

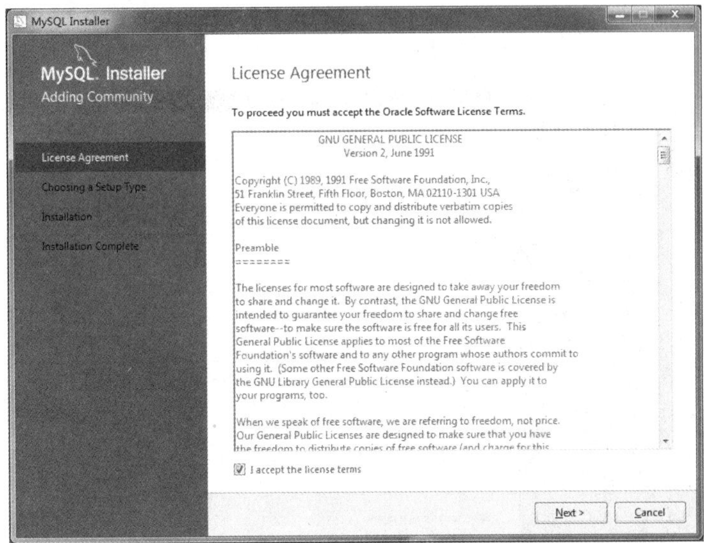

图 4-4

（4）选择安装类型。第一个选项包含一些 MySQL 其他组件，如果只安装 MySQL 数据库，选择第二项 Server only 就行。这里直接选择默认的 Developer Default 选项，如图 4-5 所示。

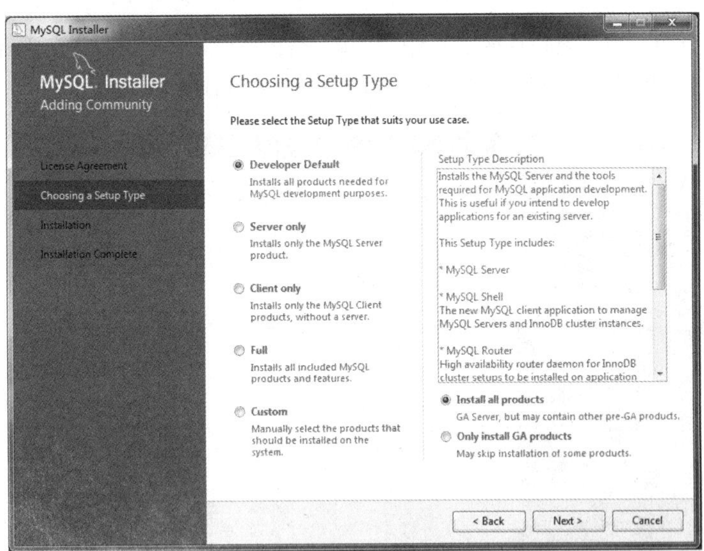

图 4-5

（5）检查必需项（Check Requirements）。选择一个选项，然后单击下面的 Check 选项，如果有弹窗说明该软件没有安装需求的版本或者额外组件，如果已经安装了，则前面会多一个绿色的钩，说明可以使用。

如果没有达到要求，则需要手动安装额外的软件。

如果有些产品不需要用的话，不需要安装额外软件，直接单击 Next 按钮就可以了。此时会弹出一个窗口，忽略它，直接单击 Yes 按钮就可以了。

（6）等待安装完成，然后一直单击 Next 按钮。

（7）如图 4-6 所示，有 3 个使用类型——开发者、服务器、网络专用服务器，根据个人需求选择。如果是个人，一般选择开发者就可以了。MySQL 的 TCP/IP 默认端口都是 3306。

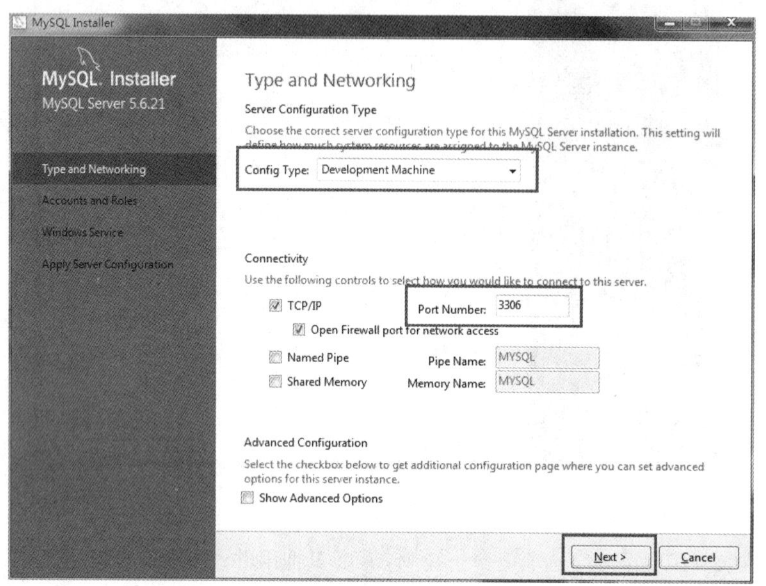

图 4-6

（8）图 4-7 上面设置的是最高权限密码，这个密码很重要，务必设置一个不容易被破解的。下面是用户设置，这个在以后添加删除用户也是可以的，安装的时候可以忽略。

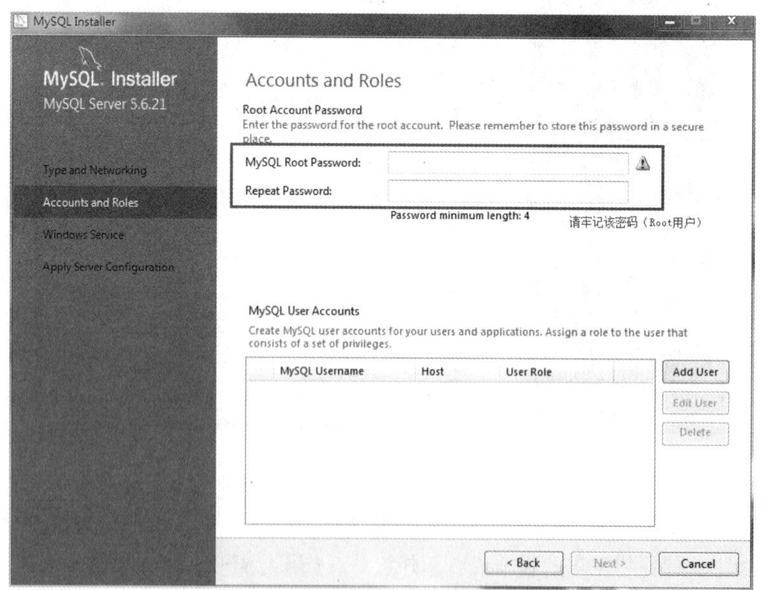

图 4-7

（9）图 4-8 中是 Windows 系统服务和插件扩展的选项，剩下的都是一些检查或者开启状态之类的（见图 4-9），这里默认一直单击 Next 按钮进入下一步。

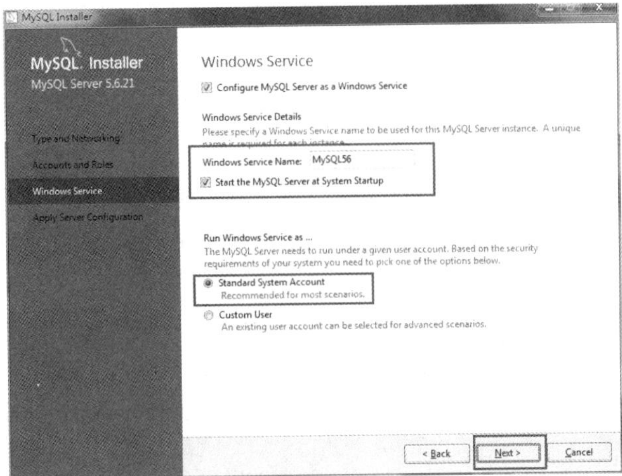

图 4-8

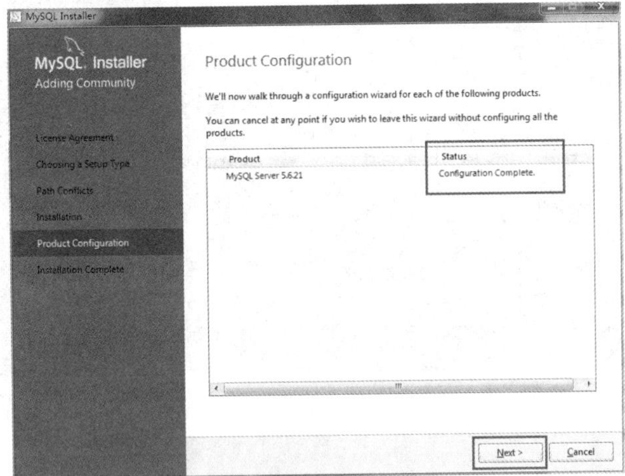

图 4-9

（10）安装完成，如图 4-10 所示。

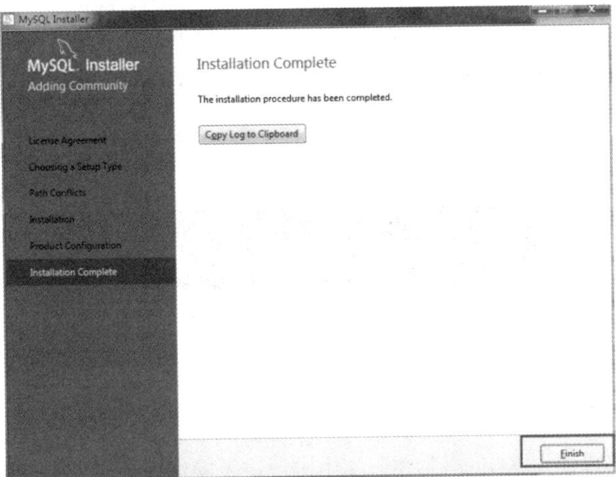

图 4-10

2. 安装成功验证

（1）打开命令行窗口，如图 4-11 所示。

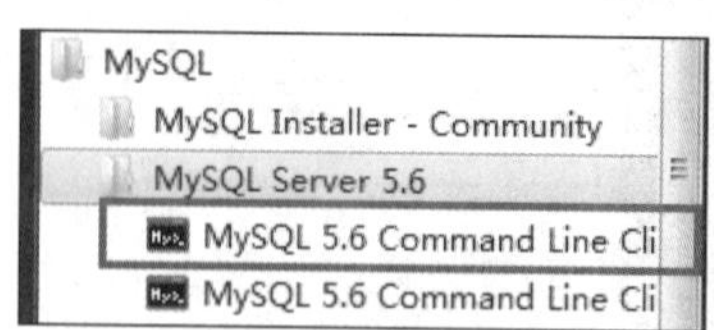

图 4-11

（2）输入 Root 密码，并按回车。

（3）如图 4-12 所示，输入显示所有数据库命令：show databases;，一定要有分号，并按回车键。将会显示系统默认的 4 个数据库。

```
MySQL 5.6 Command Line Client - Unicode
Server version: 5.6.21-log MySQL Community Server (GPL)

Copyright (c) 2000, 2014, Oracle and/or its affiliates. All rights reserved.

Oracle is a registered trademark of Oracle Corporation and/or its
affiliates. Other names may be trademarks of their respective
owners.

Type 'help;' or '\h' for help. Type '\c' to clear the current input statement.

mysql> show databases;
+--------------------+
| Database           |
+--------------------+
| information_schema |
| mysql              |
| performance_schema |
| test               |
+--------------------+
4 rows in set (0.00 sec)

mysql>
```

图 4-12

（二）建立数据库

启动 MySQL 监视器后就可以使用 SQL 语句创建数据库、建表等操作，也可以下载相应的图形界面的 MySQL 管理工具进行相应的操作，这些 MySQL 管理工具有免费的，也有需要购买的。本节主要任务是怎样在 JSP 中和 MySQL 建立连接，所以采用在 MS-DOS 命令行窗口输入 SQL 语句来建立数据库和创建表。MySQL 要求 SQL 语句必须用 “;” 号结束。在编辑 SQL 语句的过程中可以使用\c 终止当前 SQL 语句的编辑。

下面使用 MySQL 监视器创建一个名字为 studb 的数据库，如图 4-13 所示。

图 4-13

（三）MySQL 数据库的基本使用

创建数据库后就可以使用 SQL 语句在该库中进行建表操作，为了在某个数据库中建表，必须首先进入该数据库，命令为：user 数据库名。

进入数据库 studb 的操作如图 4-14 所示。

```
mysql> use studb
Database changed
mysql>
```

图 4-14

下面建立一个名字为 student 的表，建表过程如图 4-15 所示。

```
mysql> create table student(
    -> number varchar(20) not null primary key,
    -> name varchar(20) not null,
    -> birthday date,
    -> stuclass varchar(50)
    -> );
Query OK, 0 rows affected (0.01 sec)
```

图 4-15

建表之后，就可以使用 SQL 语句对表进行添加、更新和查询操作。现在向表 student 中添加 2 条记录，并使用 SQL 语句查询添加到表中的记录。操作过程如图 4-16 和图 4-17 所示。

```
mysql> insert into student values('001','张三','1988-10-10','软件S16-1班');
Query OK, 1 row affected (0.00 sec)

mysql> insert into student values('002','李四','1988-05-10','软件S16-2班');
Query OK, 1 row affected (0.00 sec)
```

图 4-16

```
mysql> select * from student;
+--------+------+------------+-------------+
| number | name | birthday   | stuclass    |
+--------+------+------------+-------------+
| 001    | 张三 | 1988-10-10 | 软件S16-1班 |
| 002    | 李四 | 1988-05-10 | 软件S16-2班 |
+--------+------+------------+-------------+
2 rows in set (0.00 sec)
```

图 4-17

任务三　连接 MySQL 数据库

任务要求

本任务要求掌握加载 JDBC 数据库驱动的步骤，通过程序能够连接到数据库。

任务实现

从图 4-1 中可以看出，对于 MySQL 数据库，只要安装 MySQL 驱动，JDBC 就可以不关心具体的连接过程而对 MySQL 进行操作；如果是 SQL Server，只需要安装 SQL Server 驱动，JDBC 就可以不关心具体的连接过程而对 SQL Server 进行操作。

从这里可以看出，要连接到不同厂商的数据库，应该首先安装相应厂商的数据库驱动。这就是数据库连接的其中一种方式——数据库厂商驱动连接。

（一）加载 JDBC 数据库驱动

由于是数据库厂商自己提供的专属驱动程序，因此这类驱动程序的弹性较差，往往只适用于自己的数据库系统，甚至只适合某个版本的数据库系统。如果后台数据库换了一个或者版本升级了，则就有可能需要更换数据库驱动程序。

使用厂商驱动，有下列两个步骤。

（1）到相应的数据库厂商网站下载厂商驱动，或者从数据库安装目录下找到相应的厂商驱动包，复制到 Web 项目的 WEB-INF/lib 下。以 MySQL 为例，可以将“mysql-connector-java-x.x.x(版本号)-bin.jar”复制到 Web 项目的 WEB-INF/lib 下。右键单击 MySQL 的驱动 jar 包，把它加入到工程。

（2）在 JDBC 代码中，设定特定的驱动程序名称和 url。不同的驱动程序和不同的数据库可以采用不同驱动程序名称和 url。

（二）建立数据库连接

JDBC 通过把特定数据库厂商数据库操作的细节抽象，得到一组类和接口，这些类和接口包含在 java.sql 包中，这样，就可以被任何具有 JDBC 驱动的数据库使用，从而实现数据库访问功能的通用化。

DriverManager 类是 JDBC 的管理层，作用于用户和驱动程序之间。它跟踪可用的驱动程序，并在数据库和相应驱动程序之间建立连接。该类负责加载、注册 JDBC 驱动程序，管理应用程序和已注册的驱动程序的连接。DriverManager 类的常用方法见表 4-1。

表 4-1　DriverManager 类的常用方法

方法名	功能说明
Static connection getConnection(String url, String user, String password)	用于建立到指定数据库 URL 的连接。其中，url 为 jdbc:subprot6col:subname 形式的数据库 url；user 为数据库用户名；password 为用户的密码
static Driver getDriver(String url)	用于返回能够打开 url 所指定的数据库的驱动程序

对于简单的应用程序，只需要直接使用该类的方法 DriverManager. getConnection 进行连接即可。通过调用方法 Class.forName，将显式地加载驱动程序类。使用 JDBC-ODBC 桥驱动程序建立连接的语句如下：

```
String dbName = "test";
String userName = "root";
String userPasswd = "123456";
String url = "jdbc:mysql://localhost/" + dbName + "?user=" + userName + "&password=" + userPasswd;
Class.forName("com.mysql.jdbc.Driver");
Connection conn = DriverManager.getConnection(url);
```

Connection 接口代表与数据库的连接，并拥有创建 SQL 语句的方法，以完成基本的 SQL 操作，同时为数据库事务处理提供提交和回滚的方法。一个应用程序可与单个数据库有一个或多个连接，也可以与多个数据库有连接。

任务四　查询数据操作

任务要求

本任务要求掌握通过 JDBC 技术对数据进行正确查询的方法。

任务实现

和数据库建立连接后，就可以使用 JDBC 提供的 API 和数据库交互信息，比如查询、修改和更新数据库中的表等。JDBC 提供的 API 可以将标准的 SQL 语句发送给数据库，实现和数据库的交互。

对一个数据库中的表进行查询操作的具体步骤如下：

1. 创建 Statement 对象

建立了与特定数据库的连接之后，就可用该连接发送 SQL 语句。Statement 对象用 Connection 的方法 createStatement 创建，如下列代码段中所示：

```
Statement stmt = conn.createStatement();
```

Statement 是 Java 执行数据库操作的一个重要接口，用于在已经建立数据库连接的基础上，向数据库发送要执行的 SQL 语句。实际上有三种 Statement 对象，Statement、PreparedStatement 和 CallableStatement。它们都专用于发送特定类型的 SQL 语句：Statement 对象用于执行不带参数的简单 SQL 语句；PreparedStatement 对象用于执行带参数或不带参数的预编译 SQL 语句；CallableStatement 对象用于执行对数据库存储过程的调用。

2. 向数据库发送 SQL 查询语句

为了执行 Statement 对象，发送到数据库的 SQL 语句将被作为参数提供给 Statement 的方法：

```
ResultSet rs = stmt.executeQuery("SELECT a, b, c FROM Table2");
```

3. 处理查询结果

Statement 对象可以调用相应的方法实现对数据库中表的查询和修改，并将查询结果存放在一个 ResultSet 类声明的对象中。也就是说 SQL 查询语句对数据库的查询操作将返回一个 ResultSet 对象。ResultSet 对象是以统一形式的列组织的数据行组成。

ResultSet 对象读取数据的方法主要是 getXXX()，参数可以使用整型表示第几列（索引号从 1 开始），还可以是列名，返回的是对应的 XXX 类型的值。如果对应那列是空值，XXX 是对象的话返回 XXX 型的空值，如果 XXX 是数字类型，如 Float 等则返回 0，boolean 则返回 false。使用 getString() 可以返回所有的列的值，不过返回的都是字符串类型的。XXX 可以代表的类型有：基本的数据类型，如整型（int）、布尔型（Boolean）、浮点型（Float，Double）、比特型（byte）等，还包括一些特殊的类型，如日期类型（java.sql.Date）、时间类型（java.sql.Time）、时间戳类型（java.sql.Timestamp）、大数型（BigDecimal 和 BigInteger）等。

注意

当使用 ResultSet 对象的 getXXX 方法查看一行记录时，不可以颠倒字段的顺序。

4. 关闭 Statement 对象

Statement 对象将由 Java 垃圾收集程序自动关闭。而作为一种好的编程风格，应在不需要 Statement 对象时显式地关闭它们。这将立即释放 DBMS 资源，有助于避免潜在的内存问题。

（一）整表数据查询

通过上面的学习了解到，查询到的结果放入 ResultSet 中，而 ResultSet 实际上是一个小表格。取数据之前，首先要介绍游标的概念（注意，不是数据库中的游标）。

游标是在 ResultSet 中可以移动的一个指针，它指向一行数据。初始时指向第 1 行的前一行，rs.next()可以将游标移到下一行，其返回值是一个布尔类型，即如果下一行有数据则返回值为 true，否则为 flase。所以必须要运行一次 next 函数之后，才能从新开始取数据，如果强行取数据的话，则会因找不到该列而报错。很明显，可以使用 rs.next()，配上 while 循环来对结果进行遍历。

【例 4-1】 编写 Web 应用程序，在计算机屏幕上用表格的形式，显示所有学生的信息，效果如图 4-18 所示。

```
<%@page import="java.sql.*" %>
<head>
    <title>整表查询数据</title>
    <%! ResultSet rs;%>
    <%
        String dbName = "studb";
        String userName = "root";
        String userPasswd = "123456";
        String url = "jdbc:mysql://localhost/" + dbName + "?user=" + userName + "&password=" +
userPasswd;
        Class.forName("com.mysql.jdbc.Driver");
        Connection conn = DriverManager.getConnection(url);
        Statement statement = conn.createStatement();
        String sql = "select * from student";
        rs = statement.executeQuery(sql);
    %>
</head>
<body>
<table border="1">
    <th colspan="4" style=" font-size: x-large; alignment: center"> 学生信息表</th>
    <tr style=" text-align: center">
        <td>学号</td>
        <td>姓名</td>
        <td>出生日期</td>
        <td>所在班级</td>
```

```
    </tr>
    <% while (rs.next()) {    %>
    <tr style=" text-align: center">
      <td><%= rs.getString(1)    %></td>
      <td><%= rs.getString(2) %> </td>
      <td><%= rs.getString("birthday")%></td>
      <td><%= rs.getString("stuclass")%></td>
    </tr>
    <%    }    %>
<%
rs.close();
statement.close();
  conn.close();
%>

</table>
</body>
```

学生信息表			
学号	姓名	出生日期	所在班级
001	张三	1988-10-10	软件S16-1班
002	李四	1988-05-10	软件S16-2班
003	蔡梦洁	1988-08-03	软件S16-2班
004	张华	1989-01-01	软件S16-3班
005	张鹏飞	1989-03-08	软件S16-3班

图 4-18

（二）条件查询

在 SQL 中 WHERE 子句用于规定选择的标准。如需有条件地从表中选取数据，可将 WHERE 子句添加到 SELECT 语句。

语法：

```
SELECT 列名称 FROM 表名称 WHERE 列 运算符 值
```

表 4-2 中的运算符可在 WHERE 子句中使用。

表 4-2　WEHER 子句中可用的运算符

运算符	描述
=	等于
<>	不等于
>	大于
<	小于
>=	大于等于
<=	小于等于
BETWEEN	在某个范围内
LIKE	搜索某种模式

注意

在某些版本的 SQL 中，运算符<>可以写为!=。

【例 4-2】 编写 Web 应用程序，客户通过 JSP 页面输入查询条件，并用表格的形式，显示学生的信息。

index.jsp 页面的代码如下：

```
<form action="first.jsp"   method="post" >
<table>
  <th colspan="2">查询条件</th>
  <tr>
    <td>学号：</td>
    <td><input type="text" name="stuid" id="stuid"/> </td>
  </tr>
  <tr aria-rowspan="2">
    <td><input type="submit" value="提交"></td>
  </tr>
</table>
</form>
```

first.jsp 页面的代码如下：

```
<%@page import="java.sql.*" %>
<% String id=request.getParameter("stuid");%>

<head>
    <title>条件查询</title>
    <%! ResultSet rs;%>
    <%
            String dbName = "studb";
            String userName = "root";
            String userPasswd = "123456";
            String url = "jdbc:mysql://localhost/" + dbName + "?user=" + userName + "&password=" +
userPasswd;
            Class.forName("com.mysql.jdbc.Driver");
            Connection conn = DriverManager.getConnection(url);
            Statement statement = conn.createStatement();
            String sql = "select * from student where number ='"+id+"'";
            rs = statement.executeQuery(sql);
     %>
</head>
<html>

<body>
<table border="1">
```

```
    <th colspan="4" style=" font-size: x-large; alignment: center"> 学生信息表</th>
    <tr style=" text-align: center">
        <td>学号</td>
        <td>姓名</td>
        <td>出生日期</td>
        <td>所在班级</td>
    </tr>
    <% while (rs.next()) {   %>
    <tr style=" text-align: center">
        <td><%= rs.getString(1)   %></td>
        <td><%= rs.getString(2) %> </td>
        <td><%= rs.getString("birthday")%></td>
        <td><%= rs.getString("stuclass")%></td>
    </tr>
    <%   }   %>
<%
rs.close();
statement.close();
    conn.close();
%>

</table>
</body>
```

运行效果如图 4-19 所示。

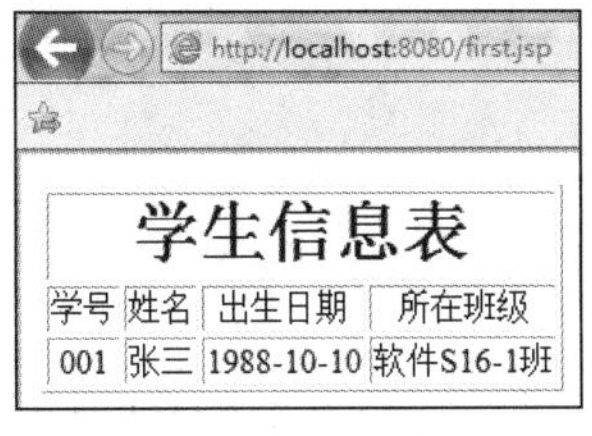

图 4-19

如果想要更进一步检索满足多个条件下的序列的话，可以使用 AND 和 OR 运算符对记录进行过滤。

在上面的例题中，想要查询出该班级下的指定学号的学生信息，只需要把对应的 SQL 语句修改成：

```
String sql = "select * from student where stuclass='软件S16-2班' and number ='"+id+"'";
```

以下语句能查询出在该班级的学生信息或者指定学号的学生信息：

```
String sql = "select * from student where stuclass='软件S16-2班' or number ='"+id+"'";
```

（三）排序查询

可以在 SQL 语句中使用 ORDER BY 语句对结果集进行排序。ORDER BY 语句默认按照升序对记录进行排序。如果希望按照降序对记录进行排序，可以使用 DESC 关键字。

【例 4-3】 编写 Web 应用程序，按照学生的班级降序排列，班级相同的按照学号升序排序，并用表格的形式，显示学生的信息。

```
<%@ page contentType="text/html;charset=UTF-8" language="java" %>
<%@page import="java.sql.*" %>
<head>
  <title>排序查询</title>
  <%! ResultSet rs;%>
  <%
      String dbName = "studb";
      String userName = "root";
      String userPasswd = "123456";
      String url = "jdbc:mysql://localhost/" + dbName + "?user=" + userName + "&password=" +
userPasswd;
      Class.forName("com.mysql.jdbc.Driver");
      Connection conn = DriverManager.getConnection(url);
      Statement statement = conn.createStatement();
      String sql = "select * from student ORDER BY stuclass DESC,number ASC   ";
      rs = statement.executeQuery(sql);
  %>
</head>
<body>
<table border="1">
  <th colspan="4" style=" font-size: x-large; alignment: center"> 学生信息表</th>
  <tr style=" text-align: center">
    <td>学号</td>
    <td>姓名</td>
    <td>出生日期</td>
    <td>所在班级</td>
  </tr>
  <% while (rs.next()) {   %>
  <tr style=" text-align: center">
    <td><%= rs.getString(1)   %></td>
    <td><%= rs.getString(2) %> </td>
    <td><%= rs.getString("birthday")%></td>
    <td><%= rs.getString("stuclass")%></td>
  </tr>
  <%   }   %>
<%
rs.close();
statement.close();
   conn.close();
%>

</table>
</body>
```

运行效果如图 4-20 所示。

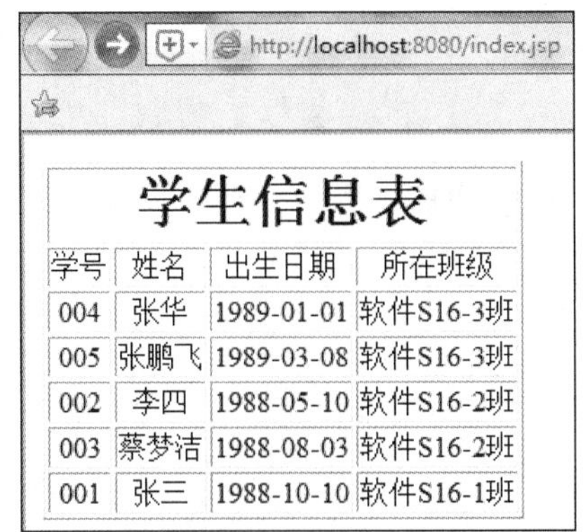

学生信息表			
学号	姓名	出生日期	所在班级
004	张华	1989-01-01	软件S16-3班
005	张鹏飞	1989-03-08	软件S16-3班
002	李四	1988-05-10	软件S16-2班
003	蔡梦洁	1988-08-03	软件S16-2班
001	张三	1988-10-10	软件S16-1班

图 4-20

任务五　增加数据操作

任务要求

本任务要求掌握通过 JDBC 技术对数据进行添加的方法。

任务实现

通过前面的学习知道，Statement 接口中 executeQuery 这个方法用来执行 SELECT 语句，它几乎是使用最多的 SQL 语句。Statement 接口中还有一个 executeUpdate 方法，可以使用该方法执行 INSERT、UPDATE 或 DELETE 语句以及 SQL DDL（数据定义语言）语句，例如 CREATE TABLE 和 DROP TABLE。INSERT、UPDATE 或 DELETE 语句的效果是修改表中零行或多行中的一列或多列。executeUpdate 的返回值是一个整数，指示受影响的行数（即更新计数）。对于 CREATE TABLE 或 DROP TABLE 等不操作行的语句，executeUpdate 的返回值总为零。

【例 4-4】 编写 Web 应用程序，用户通过 JSP 页面添加学生信息，并以表格的形式，显示学生的信息。

index.jsp 页面的代码如下：

```
<%@ page contentType="text/html;charset=UTF-8" language="java" %>
<body>
<form action="first.jsp"   method="post" >
  <table>
    <th colspan="2">增加学生信息</th>
    <tr>
       <td>学生学号：</td>
       <td><input type="text" name="stuid" id="stuid"/> </td>
    </tr>
    <tr>
```

```
        <td>学生姓名：</td>
        <td><input type="text" name="stuname" id="stuname"/> </td>
      </tr>
      <tr>
        <td>出生日期：</td>
        <td><input type="text" name="stubirthday" id="stubirthday"/> </td>
      </tr>
      <tr>
        <td>所在班级：</td>
        <td><input type="text" name="stuclass" id="stuclass"/> </td>
      </tr>
      <tr aria-rowspan="2">
        <td><input type="submit" value="提交"></td>
      </tr>
    </table>
  </form>
  </body>
  </html>
```

first.jsp 页面的代码如下：

```
  <%@ page contentType="text/html;charset=UTF-8" language="java" %>
  <%@page import="java.sql.*" %>
  <%request.setCharacterEncoding("utf-8");%>
  <% String id=request.getParameter("stuid");%>
  <% String stuName=   request.getParameter("stuname");%>
  <% String stuBirthday=request.getParameter("stubirthday");%>
  <% String stuClass=request.getParameter("stuclass");%>
  <head>
      <title>条件查询</title>
      <%! ResultSet rs;%>
      <%
            String dbName = "studb";
            String userName = "root";
            String userPasswd = "123456";
            String url = "jdbc:mysql://localhost/" + dbName + "?user=" + userName + "&password=" +
userPasswd;
            Class.forName("com.mysql.jdbc.Driver");
            Connection conn = DriverManager.getConnection(url);
            Statement statement = conn.createStatement();
            String sql = "INSERT into student VALUES('"+id+"','"+stuName+"','"+stuBirthday+"',
'"+stuClass+"')";
            int   result= statement.executeUpdate(sql);
            if(result>0)
              out.print("   <script type='text/javascript'>alert('添加成功'); </script> ");
            else
              out.print("   <script type='text/javascript'>alert('添加失败'); </script> ");
```

```
            sql = "select * from student ";
            rs = statement.executeQuery(sql);
        %>
</head>
<html>
<body>
<table border="1">
    <th colspan="4" style=" font-size: x-large; alignment: center"> 学生信息表</th>
    <tr style=" text-align: center">
        <td>学号</td>
        <td>姓名</td>
        <td>出生日期</td>
        <td>所在班级</td>
    </tr>
 <% while (rs.next()) {   %>
    <tr style=" text-align: center">
        <td><%= rs.getString(1)   %></td>
        <td><%= rs.getString(2) %> </td>
        <td><%= rs.getString("birthday")%></td>
        <td><%= rs.getString("stuclass")%></td>
    </tr>
    <%   }   %>
<%
rs.close();
preparedStatement.close();
      conn.close();
%>

</table>
</body>
</html>
```

运行效果如图 4-21 所示。

http://localhost:8080/index.jsp

增加学生信息

学生学号：006

学生姓名：陆金

出生日期：1989-11-1

所在班级：软件S16-3班

提交

http://localhost:8080/first.jsp

学生信息表			
学号	姓名	出生日期	所在班级
001	张三	1988-10-10	软件S16-1班
002	李四	1988-05-10	软件S16-2班
003	蔡梦洁	1988-08-03	软件S16-2班
004	张华	1989-01-01	软件S16-3班
005	张鹏飞	1989-03-08	软件S16-3班
006	陆金	1989-11-01	软件S16-3班

图 4-21

任务六　更新数据操作

任务要求

本任务要求掌握通过 JDBC 技术对数据进行更新的方法。

任务实现

利用上面的程序能够将数据保存到数据库中，不过，在里面出现了一句复杂的代码，其中 SQL 语句的组织依赖变量，比较容易出错。

PreparedStatement 解决了这个问题。PreparedStatement 接口是 Statement 接口的子接口，它直接继承并重载了 Statement 的方法。PreparedStatement 接口有两大特点：

（1）一个 PreparedStatement 的对象中包含的 SQL 语句是预编译的，所以当需要多次执行同一条 SQL 语句时，利用 PreparedStatement 传送这条 SQL 语句可以大大提高执行效率。

（2）PreparedStatement 的对象所包含的 SQL 语句中允许有一个或多个输入参数。创建类 PreparedStatement 的实例时，输入参数用“?”代替。在执行带参数的 SQL 语句前，必须对“?”进行赋值，为了对“?”赋值，PreparedStatement 接口中增加了大量的 setXXX 方法，用于完成对输入参数赋值。PreparedStatement 接口的常用方法见表 4-3。

表 4-3　PreparedStatement 接口的常用方法

方法	功能说明
boolean execute()	在 executeQuery()对象中执行任何 SQL 语句
ResultSet executeQuery()	在 executeQuery()对象中执行 SQL 查询，并返回该查询生成的结果集
int executeUpdate()	在 PreparedStatement 对象中执行 SQL 语句，该语句必须是 INSERT、UPDATE 或 DELETE 语句，或者是 DDL 语句
void setlnt(int x，int y)	将第 x 个参数设置为 int 值
void setString(int x，String y)	将第 x 个参数设置为 String 值

注意

PreparedStatement 接口的 setXXX()方法只列出了一部分，对其他类型的数据的操作方法只需将“XXX”换成对应的数据类型名即可。

1. 创建 PreparedStatement 对象

创建一个 Preparedstatement 接口的对象只需在建立连接后，调用 Connection 接口中的 prepareStatement()创建一个 Preparedstatement 的对象。一般形式如下：

```
String sql = "update student set birthday=?,stuclass=? where number=?";
PreparedStatement preparedStatement= conn.prepareStatement(sql);
```

2. 输入参数的赋值

Preparedstatement 中提供了大量的 setXXX(int index,Object val)方法对输入参数进行赋值，根据输入参数的 SQL 类型应选用合适的 setXXX 方法，其中 index 的值从 1 开始。一般形式如下：

```
preparedStatement.setString(1,stuBirthday);
preparedStatement.setString(2,stuClass);
preparedStatement.setString(3,id);
```

3. 执行语句

执行 Preparedstatement 对象的 excuteQuery()或者 excuteUpdate()就可以完成查询或者数据的更新。

【例 4-5】 编写 Web 应用程序，用户通过 JSP 页面修改学生信息，并以表格的形式，显示学生的信息。

index.jsp 页面的代码如下：

```
<%@ page contentType="text/html;charset=UTF-8" language="java" %>
<body>
<form action="first.jsp"  method="post" >
  <table>
    <th colspan="2">修改学生信息</th>
    <tr>
      <td colspan="2">指定需要被修改的学生信息</td>
    </tr>
    <tr>
      <td>学生学号：</td>
      <td><input type="text" name="stuid" id="stuid"/> </td>
    </tr>
    <tr>
      <td colspan="2">填写需要改正的学生信息</td>
    </tr>
    <tr>
      <td>出生日期：</td>
      <td><input type="text" name="stubirthday" id="stubirthday"/> </td>
    </tr>
    <tr>
      <td>所在班级：</td>
      <td><input type="text" name="stuclass" id="stuclass"/> </td>
    </tr>
    <tr aria-rowspan="2">
      <td><input type="submit" value="提交"></td>
    </tr>
  </table>
</form>
</body>
</html>
```

first.jsp 页面的代码如下：

```
<%@ page contentType="text/html;charset=UTF-8" language="java" %>
<%@page import="java.sql.*" %>
<%request.setCharacterEncoding("utf-8");%>
<% String id=request.getParameter("stuid");%>
<% String stuBirthday=request.getParameter("stubirthday");%>
<% String stuClass=request.getParameter("stuclass");%>
<head>
    <title>修改结果</title>
    <%! ResultSet rs;%>
    <%
        String dbName = "studb";
        String userName = "root";
        String userPasswd = "123456";
        String url = "jdbc:mysql://localhost/" + dbName + "?user=" + userName + "&password=" +
userPasswd;
        Class.forName("com.mysql.jdbc.Driver");
        Connection conn = DriverManager.getConnection(url);
        String sql = "update student set birthday=?,stuclass=? where number=?";
        PreparedStatement preparedStatement= conn.prepareStatement(sql);
        preparedStatement.setString(1,stuBirthday);
        preparedStatement.setString(2,stuClass);
        preparedStatement.setString(3,id);
        int  result= preparedStatement.executeUpdate();
        if(result>0)
          out.print("  <script type='text/javascript'>alert('修改成功'); </script> ");
        else
          out.print("  <script type='text/javascript'>alert('修改失败'); </script> ");
        sql = "select * from student ";
        rs = preparedStatement.executeQuery(sql);
  %>
</head>
<html>
<body>
<table border="1">
    <th colspan="4" style=" font-size: x-large; alignment: center"> 学生信息表</th>
    <tr style=" text-align: center">
        <td>学号</td>
        <td>姓名</td>
        <td>出生日期</td>
        <td>所在班级</td>
    </tr>
 <% while (rs.next()) {   %>
    <tr style=" text-align: center">
        <td><%= rs.getString(1)   %></td>
        <td><%= rs.getString(2) %> </td>
```

```
        <td><%= rs.getString("birthday")%></td>
        <td><%= rs.getString("stuclass")%></td>
    </tr>
<%  }  %>
<%
rs.close();
preparedStatement.close();
    conn.close();
%>
</table>
</body>
</html>
```

运行效果如图 4-22 所示。

学生信息表

学号	姓名	出生日期	所在班级
001	张三	1989-10-10	软件S16-3班
002	李四	1988-06-20	软件S16-5班
003	蔡梦洁	1988-08-03	软件S16-2班
004	张华	1989-01-01	软件S16-3班
005	张鹏飞	1989-03-08	软件S16-3班
006	陆金	1989-11-01	软件S16-3班

修改学生信息

指定需要被修改的学生信息

学生学号： 002

填写需要改正的学生信息

出生日期： 1988-6-20

所在班级： 软件S16-5班

提交

图 4-22

任务七　删除数据操作

任务要求

本任务要求掌握通过 JDBC 技术对数据进行删除的方法。

任务实现

有些情况下，需要在 PreparedStatement 中设置空值，这个时候如果使用设置具体类型的方法，如 setInt(1,null)，程序会毫不留情地报出空指针异常，所以需要做的是使用 setNull(index, type)方法来代替原来的方法。该方法是将参数值赋值为 Null，其中 type 是在 java.sql.Types 中定义的 SQL 类型，例如：

```
preparedStatement.setNull(1,Types.INTEGER);
```

当参数的值很大时，可以将参数值放在一个输入流中，再通过调用下述两种方法将其赋给特定的参数，参数 length 表示输入流中字符串的长度。基本用法如下：

```
setBinaryStream(int parameterIndex, java.io.InputStream x, int length)
setAsciiStream(int parameterIndex, java.io.InputStream x, int length)
```

【例 4-6】 编写 Web 应用程序，用户通过 JSP 页面指定要删除学生的学号，并以表格的形式，显示删除之后学生的信息。

index.jsp 的代码如下：

```
<%@ page contentType="text/html;charset=UTF-8" language="java" %>
<body>
<form action="first.jsp"   method="post" >
  <table>
    <th colspan="2">删除学生信息</th>
    <tr>
      <td>学生学号：</td>
      <td><input type="text" name="stuid" id="stuid"/> </td>
    </tr>
    <tr aria-rowspan="2">
      <td><input type="submit" value="提交"></td>
    </tr>
  </table>
</form>
</body>
</html>
```

first.jsp 的代码如下：

```
<%@ page contentType="text/html;charset=UTF-8" language="java" %>
<%@page import="java.sql.*" %>
<%request.setCharacterEncoding("utf-8");%>
<% String id=request.getParameter("stuid");%>
<head>
    <title>修改结果</title>
    <%! ResultSet rs;%>
    <%
        String dbName = "studb";
        String userName = "root";
        String userPasswd = "123456";
        String url = "jdbc:mysql://localhost/" + dbName + "?user=" + userName + "&password=" +
userPasswd;
        Class.forName("com.mysql.jdbc.Driver");
        Connection conn = DriverManager.getConnection(url);
        String sql = "DELETE from student where number=?";
        PreparedStatement preparedStatement= conn.prepareStatement(sql);
        preparedStatement.setString(1,id);
        int   result= preparedStatement.executeUpdate();
        if(result>0)
          out.print("   <script type='text/javascript'>alert('删除成功'); </script> ");
        else
          out.print("   <script type='text/javascript'>alert('删除失败'); </script> ");
        sql = "select * from student ";
        rs = preparedStatement.executeQuery(sql);
  %>
</head>
<html>
```

```
<body>
<table border="1">
    <th colspan="4" style=" font-size: x-large; alignment: center"> 学生信息表</th>
    <tr style=" text-align: center">
        <td>学号</td>
        <td>姓名</td>
        <td>出生日期</td>
        <td>所在班级</td>
    </tr>
  <% while (rs.next()) {   %>
    <tr style=" text-align: center">
        <td><%= rs.getString(1)   %></td>
        <td><%= rs.getString(2) %> </td>
        <td><%= rs.getString("birthday")%></td>
        <td><%= rs.getString("stuclass")%></td>
    </tr>
    <%   }   %>
    <%
        rs.close();
        preparedStatement.close();
        conn.close();
    %>
</table>
</body>
</html>
```

运行效果如图 4-23 所示。

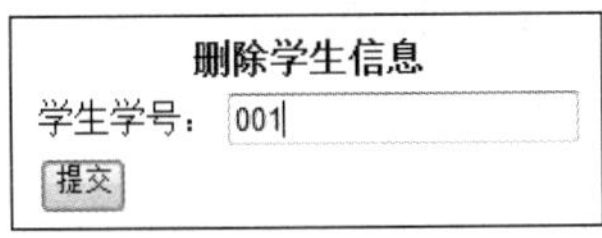

学生信息表			
学号	姓名	出生日期	所在班级
002	李四	1988-06-20	软件S16-5班
003	蔡梦洁	1988-08-03	软件S16-2班
004	张华	1989-01-01	软件S16-3班
005	张鹏飞	1989-03-08	软件S16-3班
006	陆金	1989-11-01	软件S16-3班

图 4-23

下面总结一下 PreparedStatement 相比于 Statement 有哪些优点。

（1）代码的可读性和可维护性更好。虽然从上面代码来看，用 PreparedStatement 来代替 Statement 会使代码多出几行，但这样的代码无论从可读性还是可维护性上来说，都比直接用 Statement 的代码高很多档次。

（2）PreparedStatement 尽最大可能提高性能。每一种数据库都会尽最大努力对预编译语句提供最大的性能优化，因为预编译语句有可能被重复调用，所以语句在被编译器编译后的执行代码被缓存下来，下次调用时只要是相同的预编译语句就不需要编译，只要将参数直接传入编译过的语句执行代码中就会得到执行。而 statement 语句中，即使是相同的操作，因为每次操作的数据不同，所以使整个语句相匹配的机会极小，几乎不太可能匹配。比如：

```
INSERT into student VALUES('005','张鹏飞','1989-03-08','软件S16-3班')
INSERT into student VALUES('006','陆金','1989-11-01','软件S16-3班')
```

即使是相同操作但因为数据内容不一样，所以整个语句本身不能匹配。由此可知，用 Statement 对象时，每次执行一个 SQL 命令，都会对它进行解析编译，而 PreparedStatement 对象在多次执行同一个 SQL 语句时都只解析编译一次。PreparedStatement 对象“就像一条生产线，批量生产同一型号的产品速度非常快”，这样便可极大地减少资源开销。

（3）极大地提高了安全性。传递给 PreparedStatement 对象的参数可以被强制进行类型转换，使开发人员可以确保在插入或查询数据时与底层的数据库格式匹配。这样做是为了防止有恶意的用户利用那些设计不完善的、不能正确处理字符串的应用程序来执行 SQL 注入。

任务八　常见数据库的连接

任务要求

本任务要求了解常见数据库的连接方法。

任务实现

利用驱动作为中间件的应用服务器来访问数据库的模式，其中应用服务器作为一个到多个数据库的网关，客户端通过它可以连接到不同的数据库服务器。应用服务器通常都有自己的网络协议，Java 客户端程序通过 JDBC 驱动程序将 JDBC 调用发送给应用服务器，应用服务器使用本地驱动程序访问数据库，从而完成请求，如图 4-24 所示。

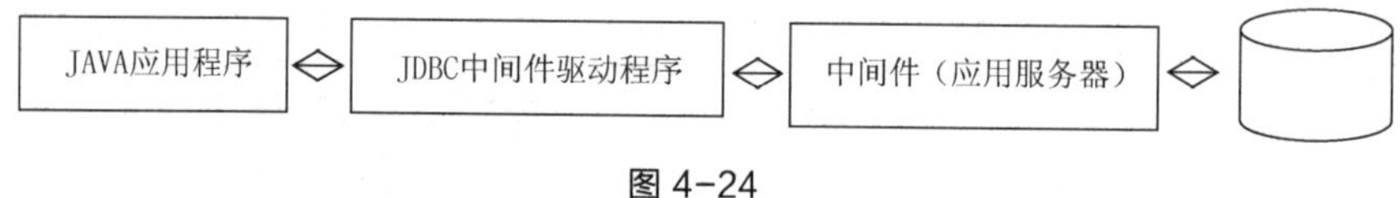

图 4-24

无论采用哪种数据库，其连接步骤和使用到的 JDBC API 都是相同的。

1. 连接 Access 数据库

（1）加载数据库驱动。

```
Class.forName("sun.jdbc.odbc.JdbcOdbcDriver");
```

（2）将数据源 student.mdb 放在项目根目录下。

```
String url="jdbc:odbc:driver={Microsoft Access Driver (*.mdb)};DBQ="+request.getRealPath("/")+
"student.mdb";
或
String spath="student.mdb";
String dbpath=application.getRealPath(spath);
String url="jdbc:odbc:driver={Microsoft Access Driver (*.mdb)};DBQ="+dbpath;
```

（3）建立连接操作。

```
Connection conn=DriverManager.getConnection(url);
```

```
Statement stmt=conn.createStatement();
```

2. 连接 SQL Server 数据库

（1）下载 Microsoft JDBC Driver 4.0 for SQL Server 文件，将下载文件中的 sqljdbc4.jar 复制后放到工程项目文件下的 WEB-INF\lib 中。

（2）右键单击 sqljdbc4.jar 文件，在弹出的框中单击 Add as Library，在弹出的框中选择默认的设置，然后添加进去。

（3）加载数据库操作。

```
String DBDRIVER="com.microsoft.jdbc.sqlserver.SQLServerDriver";
String DBURL="jdbc:jtds:sqlserver://127.0.0.1:1433/数据库名";
String DBUSER="登录名";
String PASSWORD="登录密码";
Class.forName(DBDRIVER);
```

（4）建立数据库连接。

```
Connection cn=DriverManager.getConnection(DBURL,DBUSER,PASSWORD);
Statement st=cn.createStatement();
```

3. 连接 Oracle 数据库

（1）把 Oracle 11g 的 JDBC 驱动 ojdbc6.jar 拷贝到 WEB-INF\lib\目录下。

（2）右键单击 ojdbc6.jar 文件，在弹出的框中单击 Add as Library，在弹出的框中选择默认的设置，然后添加进去。

（3）加载数据库操作。

```
String DBDRIVER="oracle.jdbc.driver.OracleDriver";
String DBURL="jdbc:oracle:thin:@localhost:1521:orcl";//orcl为数据库的SID
String DBUSER="登录名";
String PASSWORD="登录密码";
Class.forName(DBDRIVER);
```

（4）建立数据库连接。

```
Connection cn=DriverManager.getConnection(DBURL,DBUSER,PASSWORD);
Statement st=cn.createStatement();
```

任务九　拓展实训

任务要求

在实际开发中，经常需要向数据库发送多条 SQL 语句，这时如果逐条执行这些 SQL 语句，效率会很低。为此，JDBC 提供了批处理机制，即同时执行多条 SQL 语句。

任务实现

为了避免重复代码的书写，可以使用 PreparedStatement 实现批处理。与 Statement 相比，PreparedStatement 灵活许多，它既可以使用完整的 SQL，也可以使用带参数的不完整 SQL。下面通过一个案例来演示，如何使用 PreparedStatement 实现批处理。

【实训 4-1】 编写 Web 应用程序，批量导入学生的信息，并以表格的形式，显示学生的信息。

```
<%@ page import="java.sql.Connection" %>
<%@ page import="java.sql.DriverManager" %>
<%@ page import="java.sql.PreparedStatement" %>
<%@ page import="java.sql.ResultSet" %>
<%@ page contentType="text/html;charset=UTF-8" language="java" %>
<html>
  <head>
    <title>批量添加</title>
     <%! ResultSet rs;%>
<%

        String dbName = "studb";
        String userName = "root";
        String userPasswd = "123456";
        String url = "jdbc:mysql://localhost/" + dbName + "?user=" + userName + "&password=" +
userPasswd;
        Class.forName("com.mysql.jdbc.Driver");
        Connection conn = DriverManager.getConnection(url);
        //通过数据模拟成批量的数据
        String[] numbers=new  String[]{"007","008","009","010" };
        String[] names=new  String[]{ "赵博","王松","张蕾","王哲"};
        String[] birthdays=new  String[]{"1989-11-01","1989-11-02","1989-11-03","1989-11-04"};
        String[] stuclass=new  String[]{"软件S16-1班","软件S16-2班","软件S16-3班","软件S16-4班"};
        String sql = "INSERT into student VALUES( ?,?,?,?)";
        PreparedStatement preparedStatement= conn.prepareStatement(sql);
        for (int i = 0; i <4 ; i++) {
         preparedStatement.setString(1,numbers[i]);
         preparedStatement.setString(2,names[i]);
         preparedStatement.setString(3,birthdays[i]);
         preparedStatement.setString(4,stuclass[i]);
         preparedStatement.addBatch();
         }
        int[] result= preparedStatement.executeBatch();
        if(result.length>0)
          System.out.print("  <script type='text/javascript'>alert('导入信息成功'); </script> ");
        else
          System.out.print("  <script type='text/javascript'>alert('导入信息失败'); </script> ");
        sql = "select * from student ";
        rs = preparedStatement.executeQuery(sql);
  %>

  </head>
  <body>
      <table border="1">
```

```
            <th colspan="4" style=" font-size: x-large; alignment: center"> 学生信息表</th>
            <tr style=" text-align: center">
                <td>学号</td>
                <td>姓名</td>
                <td>出生日期</td>
                <td>所在班级</td>
            </tr>
            <% while (rs.next()) {   %>
            <tr style=" text-align: center">
                <td><%= rs.getString(1)   %></td>
                <td><%= rs.getString(2) %> </td>
                <td><%= rs.getString("birthday")%></td>
                <td><%= rs.getString("stuclass")%></td>
            </tr>
            <%   }   %>
            <%
                rs.close();
                preparedStatement.close();
                conn.close();
            %>
        </table>
    </body>
</html>
```

运行效果如图 4-25 所示。

学生信息表			
学号	姓名	出生日期	所在班级
002	李四	1988-06-20	软件S16-5班
003	蔡梦洁	1988-08-03	软件S16-2班
004	张华	1989-01-01	软件S16-3班
005	张鹏飞	1989-03-08	软件S16-3班
006	陆金	1989-11-01	软件S16-3班
007	赵博	1989-11-01	软件S16-1班
008	王松	1989-11-02	软件S16-2班
009	张蕾	1989-11-03	软件S16-3班
010	王哲	1989-11-04	软件S16-4班

图 4-25

本章小结

本章基于 JDBC 技术，首先讲解了 MySQL 数据库的安装以及使用，其次重点讲解了 JDBC 的应用，其主要作用是封装对数据库的各种操作。

JDBC 由类和接口组成，使用 Java 开发数据库应用都需要 4 个主要的接口：Driver、Connection、Statement、ResultSet，这些接口定义了使用 SQL 访问数据库的一般架构。

JDBC 的操作步骤如下。

（1）加载驱动程序。驱动程序由各个数据库生产商提供。

（2）连接数据库。连接时要提供连接路径、用户名、密码。

（3）实例化操作。通过连接对象实例化 Statement 或 PreparedStatement 对象。

（4）操作数据库。使用 Statement 或 PreparedStatement 操作。如果查询，则全部的查询结果使用 ResultSet 进行接收。

课后练习

1. 填空题

（1）JDBC 的主要任务是：____________、________________、______________。

（2）JSP 中常用的两种数据库连接方式是：____________和________________。

（3）使用 Statement 对象的_____________方法执行查询语句，使用______________方法执行更新语句。

2. 选择题

（1）在 JSP 中使用 JDBC 语句访问数据库，正确导入 SQL 类库的语句是（　　）。（选择一项）

A. <%@ page import="java.sql.*" %>　　B. <%@ page import="sql.*" %>

C. <% page import="java.sql.*" %>　　D. <%@ import="java.sql.*" %>

（2）在 JDBC 中，负责执行 SQL 语句的接口是（　　）。（选择两项）

A. Connection　　B. Statement

C. Result　　D. PreparedStatement

（3）在 JDBC 中，用于封装查询结果的是（　　）。（选择一项）

A. ResultSet　　B. Connection

C. PreparedStatement　　D. DriverManager

3. 简答题

（1）简述什么是 JDBC。

（2）简述 JDBC 的实现步骤。

4. 编程题

（1）完成某网站的用户注册功能，通过数据库实现将用户注册信息保存到数据库中，同时要求用户名不能重复。

（2）完成某网站的用户登录功能，通过数据库验证用户名和密码的合法性。

PART05

项目五

JavaBean技术的应用

JavaBean 是使用 Java 语言开发的一个可重用的组件，在 JSP 的开发中可以使用 JavaBean 减少重复代码，使整个 JSP 代码的开发更简洁。JavaBean 提高了程序的可读性、易维护，而且提高了代码的重用性，节省了开发时间。

课堂学习目标

- 了解 JavaBean 技术
- 了解 JavaBean 的规则
- 掌握 JavaBean 的应用

素养拓展

- 近朱者赤，近墨者黑

素养拓展

任务一　JavaBean 技术简介

任务要求

本任务要求了解 JavaBean 概述、JavaBean 的种类。

任务实现

（一）JavaBean 概述

在 JSP 网页开发的初级阶段，没有逻辑分层概念的时候，需要把 Java 代码直接嵌入到网页中，然后利用 Java 代码对页面中的一组业务逻辑处理，开发流程如图 5-1 所示。

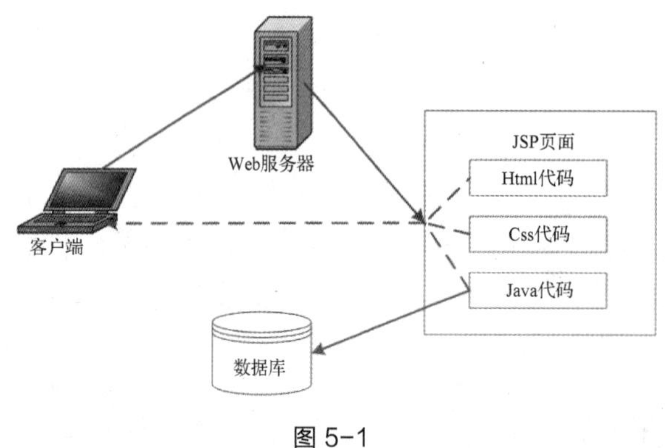

图 5-1

这样的开发虽然流程简单，但是 Java 代码嵌入到 JSP 页面中给修改和维护带来很多麻烦，因为 JSP 页面中包含很多 HTML、CSS 和 JavaScript 等页面前端代码，如果再加入后台业务逻辑代码，不利于技术分层实现，也不能体现面向对象的优势，出现代码无法重用的情况。

如果使前端 HTML 代码和后端 Java 代码分离，把实现业务逻辑的类单独封装，然后在 JSP 页面中调用，就可以降低 HTML 和 Java 代码的耦合度，使 JSP 页面简洁，易于重用和维护，这样的类就是一个 JavaBean 组件类。在 JavaWeb 应用中加入 JavaBean 的开发模式如图 5-2 所示。

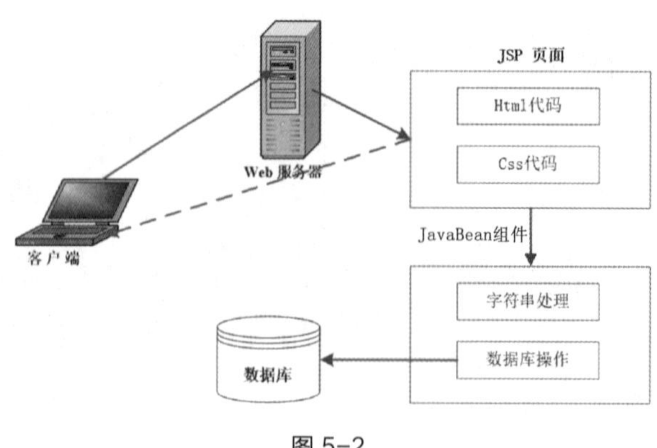

图 5-2

（二）JavaBean 的种类

JavaBean 一般分为可视化组件和非可视化组件两种。可视化组件可以是简单的 GUI 元素，如按钮或文本框，也可以是复杂的，如报表组件；非可视化组件没有 GUI 表现形式，用于封装业务逻辑、数据库操作等。JavaBean 最大的优点在于可以实现代码的可重用性。而现在，JavaBean 更多地应用于非可视化领域，同时，JavaBean 在服务器端的应用也表现出强大的优势。非可视化的 JavaBean 可以很好地实现业务逻辑、控制逻辑和显示页面的分离，现在多用于后台处理，使得系统具有更好的健壮性和灵活性。JSP + JavaBean 和 JSP + JavaBean + Servlet 成为当前开发 Web 应用的主流模式。下面通过一个实例来了解一下非可视化的 JavaBean。

【例 5-1】 创建名称为 User 的 JavaBean 对象，用于封装用户名和密码，该类位于 jspSamples.unit5.javaBeanSamples 包中。

```
package jspSamples.unit5.javaBeanSamples;
public class User {

    private String username = null;
    private String password = null;
    public String getUsername() {
        return username;
    }
    public void setUsername(String username) {
        this.username = username;
    }
    public String getPassword() {
        return password;
    }
    public void setPassword(String password) {
        this.password = password;
    }
    public User() {
    }
}
```

提示

在 JavaBean 的规范中，要求 JavaBean 对象提供默认无参数的构造方法。除了默认无参数的构造方法外，JavaBean 对象也可以根据相应属性提供构造方法。

任务二　JavaBean 的规则

任务要求

本任务要求了解 JavaBean 编写规范、JavaBean 编写要求、JavaBean 命名规范、JavaBean 的包、JavaBean 的结构。

任务实现

（一）JavaBean 编写规范

JavaBean 实际上是按照技术标准所制定的命名和设计规范编写的 Java 类。这些类遵循一个接口格式，以便于使方法命名、底层行为，其最大的优点在于可以实现代码的可重用性。Bean 并不需要继承特别的基类（BaseClass）或实现特定的接口（Interface）。Bean 的编写规范是 Bean 的容器（Container）能够分析一个 Java 类文件，并将其方法（Methods）翻译成属性（Properties），即把 Java 类作为一个 Bean 类使用。Bean 的编写规范包括 Bean 类的构造方法、定义属性和访问方法编写规则。

（二）JavaBean 编写要求

（1）所有的 JavaBean 必须放在一个包（Package）中。

（2）JavaBean 必须生成 public class 类，文件名称应该与类名称一致。

（3）所有属性必须封装，一个 JavaBean 类不应有公共实例变量，类变量都为 private 的。

（4）属性值应该通过一组存取方法（getXxx 和 setXxx）来访问。对于每个属性，应该有一个带匹配公用 getter 和 setter 方法的专用实例变量。

（5）JavaBean 类必须有一个空的构造函数。类中必须有一个不带参数的公用构造器，此构造器也应该通过调用各个属性的设置方法来设置属性的默认值。

（三）JavaBean 命名规范

JavaBean 的命名规范如下。

（1）包命名：全部字母小写。

（2）类命名：每个单词首字母大写。

（3）属性名：第一个单词全部小写，之后每个单词首字母大写。

（4）方法名：与属性命名方法相同。

（5）常量名：全部字母大写。

（四）JavaBean 的包

包即 package，JavaBean 的包和前面章节中介绍的包的含义基本上是一样的，但是也有区别。前面介绍的包都是 Java 本身定义的，而 JavaBean 的包是用户自己定义的。

每一个 JavaBean 源文件被编译成.class 文件后，都必须存放在相应的文件夹下，存放这个.class 文件的文件夹就是一个包。JavaBean 的包必须存放在特定的目录下，在每个 JSP 引擎中都规定了存放 JavaBean 包的位置，不同的 JSP 引擎对 JavaBean 的存放位置有不同的规定，如在 Tomcat 中，JavaBean 的所有包都存放在 WEB-INF/classes 文件夹中。如果存在多级目录，则需要将.class 文件所在目录的所有上级目录包含到包名称中，每一级目录之间用英文标点“.”隔开。例如下面的代码：

```
package jsp.example.mybean;
```

（五）JavaBean 的结构

属性：即 JavaBean 类的成员变量，用于描述 JavaBean 对象的状态，对象属性值的改变触发事件，属性本身就是事件源。

方法：在 JavaBean 中，函数和过程统称为方法，通过方法来改变和获取属性的值。方法可以分为构造方法、访问方法和普通方法等。

事件：事件实际上是一种特殊的 JavaBean，属性值的改变触发事件，事件激发相关对象做出反应，通过 JavaBean 注册对象事件监听者机制来接收、处理事件，它实现了 JavaBean 之间的通信。

任务三　JavaBean 的应用

任务要求

本任务要求获取 JavaBean 属性、给 JavaBean 属性赋值、在 JSP 页面中使用 JavaBean、使用常见问题解决办法。

任务实现

（一）获取 JavaBean 的属性信息

在 JavaBean 类中，为了防止外部直接对 JavaBean 属性的调用，通常将 JavaBean 中的属性设置为私有的（private），但是需要为其提供公共的（public）访问方法，也就是 getXXX()。

<jsp:useBean>标签用于在指定的域范围内查找指定名称的 JavaBean 对象，如果存在则直接返回该 JavaBean 对象的引用，如果不存在则实例化一个新的 JavaBean 对象并将它以指定的名称存储到指定的域范围中。

语法：

```
<jsp:useBean id="beanName"class="package.class" scope="page|request|session|application"/>
```

其中，id 属性用于指定 JavaBean 实例对象的引用名称和其存储在域范围中的名称；class 属性用于指定 JavaBean 的完整类名（即必须带有包名）；scope 属性用于指定 JavaBean 实例对象所存储的域范围，其取值只能是 page、request、session 和 application 等四个值中的一个，其默认值是 page。

<jsp:getProperty>标签用于读取 JavaBean 对象的属性，也就是调用 JavaBean 对象的 getter 方法，然后将读取的属性值转换成字符串后插入输出的响应正文中。

语法：

```
<jsp:getProperty name="beanInstanceName" property="PropertyName" />
```

其中，name 属性用于指定 JavaBean 实例对象的名称，其值应与<jsp:useBean>标签的 id 属性值相同；property 属性用于指定 JavaBean 实例对象的属性名。

微课：显示 JavaBean 的属性信息

下面通过实例说明如何获取 JavaBean 的属性信息。

【例 5-2】 在 JSP 中显示 JavaBean 的属性信息。

创建名称为 Person 的类，该类封装 Person 对象的 JavaBean，在 Person

类中创建属性，并提供相应的 getXXX()方法。关键代码如下：

```
package jspSamples.unit5.javaBeanSamples;
/**
 * @author simple
 * Person类就是一个最简单的JavaBean
 */
public class Person {
    //------------------Person类封装的私有属性---------------------------------
    // 姓名 String类型
    private String name = "SimpleLee";
    // 性别 String类型
    private String sex = "女";
    // 年龄 int类型
    private int age = 30;
    //是否已婚 boolean类型
    private boolean married = true;
     //------------Person类的无参数构造方法---------------------------------------
    /**
     * 无参数构造方法
     */
    public Person() {
    }
     //------------Person类对外提供的用于访问私有属性的public方法-------------------
    public String getName() {
        return name;
    }
    public void setName(String name) {
        this.name = name;
    }
    public String getSex() {
        return sex;
    }
    public void setSex(String sex) {
        this.sex = sex;
    }
    public int getAge() {
        return age;
    }
    public void setAge(int age) {
        this.age = age;
    }
    public boolean isMarried() {
        return married;
    }
    public void setMarried(boolean married) {
```

```
        this.married = married;
    }
}
```

提示

在本实例中演示了如何获取 JavaBean 的属性信息，所以对 Person 类中的属性设置了默认值，可以通过 getXXX()方法获取。

在 JSP 页面中获取 JavaBean 中的属性信息，该操作通过 JSP 动作标签获取。关键代码如下：

```
<%@ page language="java" contentType="text/html; charset=GBK"   pageEncoding="GBK"%>
<!DOCTYPE html PUBLIC "-//W3C//DTD HTML 4.01 Transitional//EN" "http://www.w3.org/TR/html4/
loose.dtd">
<html>
<head>
<meta http-equiv="Content-Type" content="text/html; charset=GBK">
<title>PersonJSP</title>
</head>
<body>
 <jsp:useBean id="person" class="jspSamples.unit5.javaBeanSamples.Person" scope="page">
</jsp:useBean>
 <div>
       <ul>
             <li>
                   姓名:<jsp:getProperty property="name" name="person"/>
             </li>
             <li>
                   性别：<jsp:getProperty property="sex" name="person"/>
             </li>
             <li>
                   年龄：<jsp:getProperty property="age" name="person"/>
             </li>
             <li>
                   已婚：<jsp:getProperty property="married" name="person"/>
             </li>
       </ul>
 </div>
</body>
```

提示

在 JSP 中尽量不要出现 Java 代码，这样避免页面显示和逻辑混杂在一起，所以实例采用 JSP 的动作标签使页面整洁。

本实例中主要通过<jsp:useBean>标签实例化 Person 的 JavaBean 对象，<jsp:getProperty>标签获取 JavaBean 中的属性信息。运行效果如图 5-3 所示。

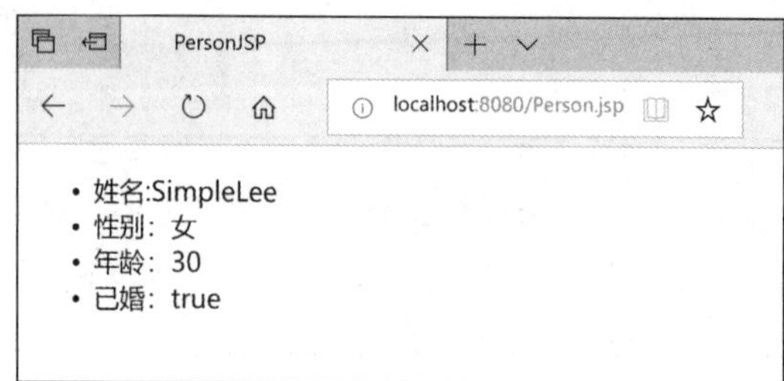

图 5-3

提示

上面在 Person.jsp 中使用<jsp:useBean id="person" class="jspSamples.unit5.javaBeanSamples.Person" scope="page"/>来实例化 person 对象的过程实际上是执行了上述的 Java 代码来实例化 Person 对象。这就是<jsp:useBean>标签的执行原理：在指定的域范围内查找指定名称的 JavaBean 对象，如果存在则直接返回该 JavaBean 对象的引用，如果不存在则实例化一个新的 JavaBean 对象并将它以指定的名称存储到指定的域范围中。

（二）对 JavaBean 属性赋值

微课：对 JavaBean 属性赋值

编写 JavaBean 对象要遵循 JavaBean 规范，JavaBean 规范中的访问器 setXXX()方法，用于对 JavaBean 中的属性赋值，如果对 JavaBean 对象提供了 setXXXX()方法，在 JSP 页面中就可以通过<jsp:setProperty>对其进行赋值。

给已经实例化的 JavaBean 对象的属性赋值，一共有四种形式，以下逐一介绍。

1. <jsp:setProperty name="JavaBean 实例名" property="*" />，跟表单关联

【例 5-3】 使用 JSP 表单自动关联 JavaBean 提交的信息。

创建含有表单的 form.jsp，包含姓名、性别、年龄信息。关键代码如下：

```
<%@ page contentType="text/html;charset=GBK" language="java" %>
<html>
<head>
    <title>Title</title>
</head>
<body>
<%request.setCharacterEncoding("GBK");%>
<form action="work.jsp" method="post">
    <table>
        <tr>
            <td>姓名:</td>
            <!--这里的name属性的值一定要和JavaBean中相应属性的名字相同-->
            <td><input type="text" name="name" value=""></td>
        </tr>
        <tr>
            <td>性别:</td>
```

```
            <td><input type="text" name="sex" value=""></td>
        </tr>
        <tr>
            <td>年龄：</td>
            <td><input type="text" name="age" value=""></td>
        </tr>
        <tr>
            <td colspan="2" align="center"><input value="提交" type="submit"></td>
        </tr>
    </table>
</form>
</body>
</html>
```

运行效果如图 5-4 所示。

图 5-4

创建含有表单的 work.jsp，通过 JavaBean 自动匹配并取得表单域值，显示在页面上。关键代码如下：

```
<%@ page contentType="text/html;charset=GBK" language="java" %>
<html>
<head>
    <title>Title</title>
</head>
<body>
<%request.setCharacterEncoding("GBK");%>
<jsp:useBean id="ps" class="jspSamples.unit5.javaBeanSamples.Person" scope="page"/>
<!--在这里因为前面form中的name、sex、age和Javabean中一样，所以进行了自动匹配，property="*",
表示匹配的属性是所有的-->
<jsp:setProperty name="ps" property="*" />
姓名:<jsp:getProperty property="name" name="ps"/><BR>
性别:<jsp:getProperty property="sex" name="ps"/><BR>
年龄:<jsp:getProperty property="age" name="ps"/><BR>
</body>
</html>
```

运行效果如图 5-5 所示。

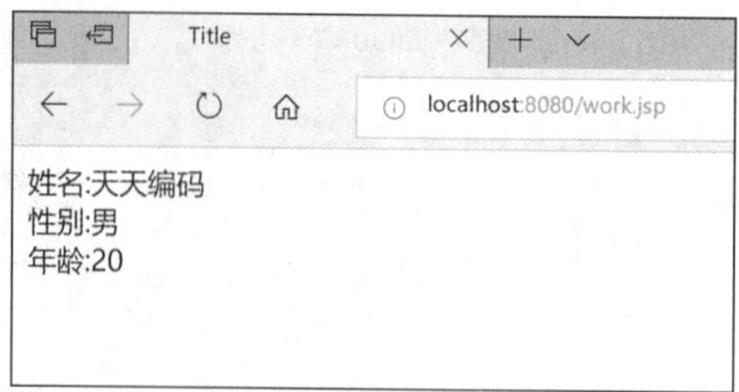

图 5-5

提示　此实例中 form.jsp 页面的 form 域里的内容自动赋值匹配给了 JavaBean 里面的各个属性，关键点是 form 页面的每个域的 name 值一定跟 Person 这个 JavaBean 的每个属性一致。并且在对 JavaBean 赋值的页面使用 property="*"。

2. <jsp:setProperty name="JavaBean 实例名" property="JavaBean 属性名" />，跟表单关联

【例 5-4】 使用 JSP 表单关联 JavaBean 提交的信息，并且指定匹配哪些属性。

修改 work.jsp，指定匹配 JavaBean 里的 age 属性，并显示在页面上。关键代码如下：

```
<%@ page contentType="text/html;charset=GBK" language="java" %>
<html>
<head>
    <title>Title</title>
</head>
<body>
<%request.setCharacterEncoding("GBK");%>
<jsp:useBean id="ps" class="jspSamples.unit5.javaBeanSamples.Person" scope="page"/>
<!--在这里因为前面form中的name、sex、age和Javabean中一样，所以进行了自动匹配，如果property="age"，那么只匹配age，所以name和sex值就为空-->
<jsp:setProperty name="ps" property="age" />
姓名:<jsp:getProperty property="name" name="ps"/><BR>
性别:<jsp:getProperty property="sex" name="ps"/><BR>
年龄:<jsp:getProperty property="age" name="ps"/><BR>
</body>
</html>
```

运行效果如图 5-6 所示。

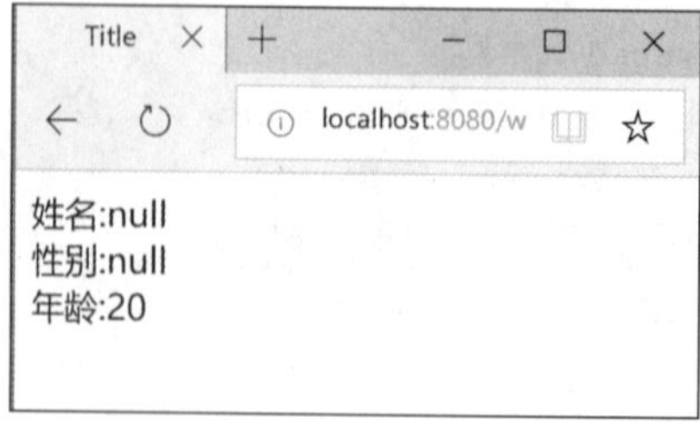

图 5-6

此实例中 form.jsp 页面的 form 域里的内容自动赋值匹配给了 JavaBean 里面的各个属性，在这里因为<jsp:setProperty name="ps" property="age" />property="age"，那么只匹配 age，所以 name 和 sex 值就为空。

3．<jsp:setProperty name="JavaBean 实例名" property="JavaBean 属性名" value="BeanValue" />，手工设置

【例 5-5】 使用 JSP 表单关联 JavaBean 提交的信息，并且手工在页面中设置每个域的值。

修改 work.jsp，指定匹配 JavaBean 里的 age 属性，并显示在页面上。关键代码如下：

```
<%@ page contentType="text/html;charset=GBK" language="java" %>
<html>
<head>
    <title>Title</title>
</head>
<body>
<%request.setCharacterEncoding("GBK");%>
<jsp:useBean id="ps" class="jspSamples.unit5.javaBeanSamples.Person" scope="page"/>
<jsp:setProperty name="ps" property="name" value="手动设置的姓名"/>
<jsp:setProperty name="ps" property="sex" value="女"/>
<jsp:setProperty name="ps" property="age"   value="22"/>
姓名:<jsp:getProperty property="name" name="ps" /><BR>
性别:<jsp:getProperty property="sex" name="ps" /><BR>
年龄:<jsp:getProperty property="age" name="ps" /><BR>
</body>
</html>
```

运行效果如图 5-7 所示。

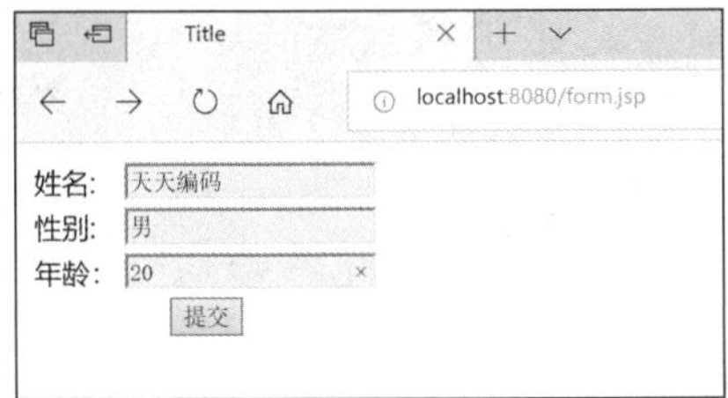

图 5-7

此实例在 work.jsp 中手工指定 value 值，如<jsp:setProperty name="ps" property="name" value="手动设置的姓名"/>，所以前一页面 form.jsp 提交的信息被设置成 JavaBean 的属性中相应的 value 值。

4．<jsp:setProperty name="JavaBean 实例名" property="propertyName" param="request 独享中的参数名" />，跟 request 参数关联

【例 5-6】 创建带参数的 JSP 表单页面，提交后使用 JavaBean 参数的设置值。

修改 form.jsp，在 action 提交 url 处传递参数。关键代码如下：

```
<form action="work.jsp?newname=SmpleLee" method="post">
    <table>
        <tr>
            <td>姓名:</td>
            <!--这里的name属性的值一定要和JavaBean中的name名字相同,age也是如此-->
            <td><input type="text" name="name" value=""></td>
        </tr>
        <tr>
            <td>性别:</td>
            <!--这里的name属性的值一定要和JavaBean中的name名字相同,age也是如此-->
            <td><input type="text" name="sex" value=""></td>
        </tr>
        <tr>
            <td>年龄：</td>
            <td><input type="text" name="age" value=""></td>
        </tr>
        <tr>
            <td colspan="2" align="center"><input value="提交" type="submit"></td>
        </tr>
    </table>
</form>
```

修改 work.jsp，使用 param 设置取得传递的参数。关键代码如下：

```
<body>
<%request.setCharacterEncoding("GBK");%>
<jsp:useBean id="ps" class="jspSamples.unit5.javaBeanSamples.Person" scope="page"/>
<jsp:setProperty name="ps" property="name" param="newname"/>
<jsp:setProperty name="ps" property="sex" />
<jsp:setProperty name="ps" property="age"  />
姓名:<jsp:getProperty property="name" name="ps" /><BR>
性别:<jsp:getProperty property="sex" name="ps" /><BR>
年龄:<jsp:getProperty property="age" name="ps" /><BR>
</body>
```

运行效果如图 5-8 所示。

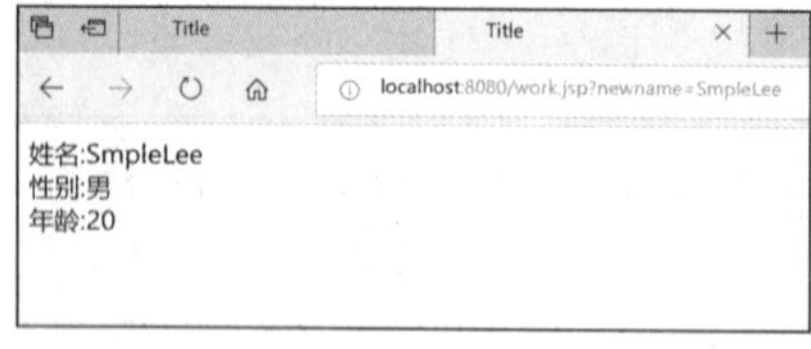

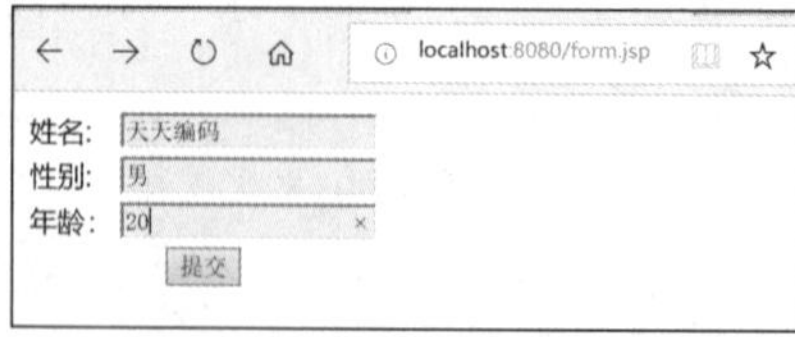

图 5-8

提示

此实例在 work.jsp 中使用 param="newname"，接收了 form.jsp 中传递的参数，改变了 Person 中的 name 属性值，实现了与 request 参数关联。

（三）JavaBean 使用中的常见问题

JavaBean 在 JSP 页面中的应用十分广泛，几乎所有的 JSP 页面中，所有的实体对象及业务逻辑的处理都有 JavaBean 封装，因此，JavaBean 与 JSP 之间的关系十分密切。JavaBean 在 JSP 页面使用中会经常遇到一些典型问题，以下逐一列举实例。

1. 解决中文乱码的 JavaBean

微课：解决中文乱码

在 JSP 页面中处理中文字符，经常会出现字符乱码的现象，特别是通过表单传递中文数据时容易产生。它的解决方法很多，如将 request 的字符集指定为中文字符集，编写 JavaBean 对中文乱码字符进行转码。

【例 5-7】 本实例要求编写对字符编码处理的 JavaBean，解决图书登记系统中图书中文信息的乱码问题。

创建名称为 Book 的类，放置在 jspSamples.unit5.javaBeanSamples 包中，实现对图书实体对象的封装。关键代码如下：

```
package jspSamples.unit5.javaBeanSamples;
public class Book {
    // 名称
    private String title;
    // 简介
    private String content;
    public String getTitle() {
        return title;
    }
    public void setTitle(String title) {
        this.title = title;
    }
    public String getContent() {
        return content;
    }
    public void setContent(String content) {
        this.content = content;
    }
```

创建对字符编码处理的 JavaBean，在此类中编写 toString()方法对字符进行编码。关键代码如下：

```
package jspSamples.unit5.javaBeanSamples;
import java.io.UnsupportedEncodingException;
public class CharactorEncoding {
 public CharactorEncoding(){
 }
 /**
  * 对字符进行转码处理
  * @param str 要转码的字符串
  * @return 编码后的字符串
  */
 public String toString(String str){
```

```
            // 转换字符
            String text = "";
            // 判断要转码的字符串是否有效
            if(str != null && !"".equals(str)){
                try {
                    // 将字符串进行编码处理
                    text = new String(str.getBytes("iso8859-1"),"GB18030");
                } catch (UnsupportedEncodingException e) {
                    e.printStackTrace();
                }
            }
            // 返回后的字符串
            return text;
    }
}
```

创建 Book.jsp 页面，用于图书信息表单登记。关键代码如下：

```
<%@ page language="java" contentType="text/html; charset=GB18030"
        pageEncoding="GB18030"%>
<!DOCTYPE html PUBLIC "-//W3C//DTD HTML 4.01 Transitional//EN" "http://www.w3.org/TR/html4/
loose.dtd">
<html>
<head>
    <meta http-equiv="Content-Type" content="text/html; charset=GB18030">
    <title>图书登记系统</title>
</head>
<body>
<form action="show.jsp" method="post">
    <table align="center" width="450" height="260" border="1">
        <tr>
            <td align="center" colspan="2" height="40" >
                <b>图书登记</b>
            </td>
        </tr>
        <tr>
            <td align="right">书  名：</td>
            <td>
                <input type="text" name="title" size="30">
            </td>
        </tr>
        <tr>
            <td align="right">简  介：</td>
            <td>
                <textarea name="content" rows="8" cols="40"></textarea>
            </td>
        </tr>
```

```
            <tr>
                <td align="center" colspan="2">
                    <input type="submit" value="登  记">
                </td>
            </tr>
        </table>
    </form>
    </body>
    </html>
```

创建 show.jsp 页面，用于对提交的图书信息表单进行处理。关键代码如下：

```
    <%@ page language="java" contentType="text/html; charset=GB18030"
                pageEncoding="GB18030"%>
    <!DOCTYPE html PUBLIC "-//W3C//DTD HTML 4.01 Transitional//EN" "http://www.w3.org/TR/html4/
loose.dtd">
    <html>
    <head>
        <meta http-equiv="Content-Type" content="text/html; charset=GB18030">
        <title>图书信息</title>
        <style type="text/css">
            #container{
                width: 450px;
                border: solid 1px;
                padding: 20px;
            }
            #title{
                font-size: 16px;
                font-weight: bold;
                color: #3399FF;
            }
            #content{
                font-size: 12px;
                text-align: left;
            }
        </style>
    </head>
    <body>
    <jsp:useBean id="book" class="jspSamples.unit5.javaBeanSamples.Book"></jsp:useBean>
    <jsp:useBean id="encoding" class="jspSamples.unit5.javaBeanSamples.CharactorEncoding">
</jsp:useBean>
    <jsp:setProperty property="*" name="book"/>
    <div align="center">
        <div id="container">
            <div id="title">
                <%= encoding.toString(book.getTitle())%>
            </div>
```

```
            <hr>
            <div id="content">
                <%= encoding.toString(book.getContent())%>
            </div>
        </div>
    </div>
</body>
</html>
```

通过<jsp:useBean>标签实例化 JavaBean 对象，如果在 JSP 页面中使用 Java 代码调用 JavaBean 对象中的属性和方法，所使用的 JavaBean 对象的变量名称为<jsp:useBean>标签中的 id 属性。

运行实例后，首先打开 Book.jsp 页面，在图书登记系统中填写正确的中文信息，如图 5-9 所示，单击“登记”按钮，其表单提交到 show.jsp 页面，运行结果如图 5-10 所示。

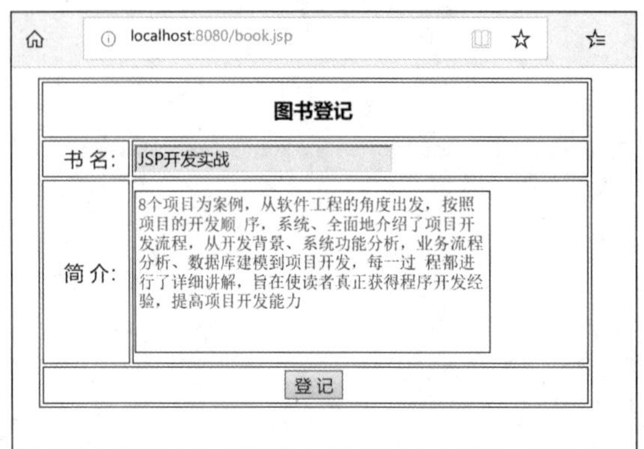

图 5-9

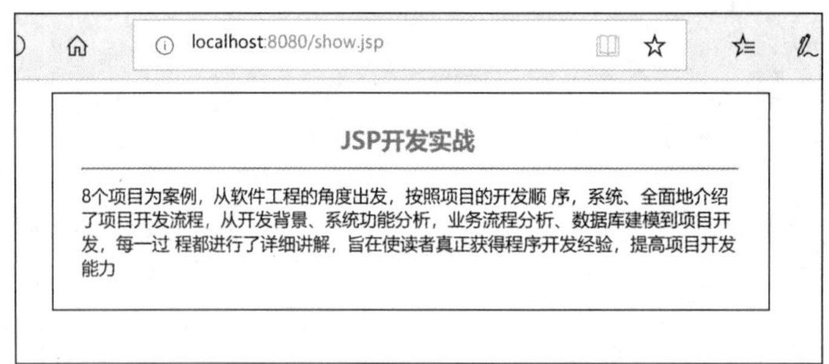

图 5-10

2. 把数组转换成字符串的 JavaBean

程序开发中，将数组转换成为字符串是经常要用到的，如表单中的复选框，在提交之后它就是一个数组对象，由于数组中的对象在业务逻辑处理中不方便，所以在实际应用过程中通常将其转换成字符串后再进行处理。

【例 5-8】 本实例要求创建将数组转换成字符串的 JavaBean，实现对“开发技术调研”表单中复选框数值的处理。

创建名称为 Paper 的类，放置在 jspSamples.unit5.javaBeanSamples 包中，实现对开发技术实体对象的封装。关键代码如下：

```
package jspSamples.unit5.javaBeanSamples;
import java.io.Serializable;
public class Paper implements Serializable {
    private static final long serialVersionUID=1L;
    private String[] languages;//定义保存编程语言的字符串数组
    private String[] technics;//定义保存掌握技术的字符串数组
    private String[] parts;//定义保存困难部分的字符数组
    public Paper(){

    }
    public String[] getLanguages() {
        return languages;
    }
    public void setLanguages(String[] languages) {
        this.languages = languages;
    }
    public String[] getTechnics() {
        return technics;
    }
    public void setTechnics(String[] technics) {
        this.technics = technics;
    }
    public String[] getParts() {
        return parts;
    }
    public void setParts(String[] parts) {
        this.parts = parts;
    }
}
```

将数组转换成字符串的 JavaBean 对象 Convert，其中 arr2Str 方法，将数组转换成指定格式的字符串。关键代码如下：

```
package jspSamples.unit5.javaBeanSamples;

public class Convert {
    /**
     * 将数组转换为字符串
     * @param arr  数组
     * @return  字符串
     */
    public String arr2Str(String[] arr){
        StringBuffer sb=new StringBuffer();
        if(arr!=null&&arr.length>0){
            for(String s:arr){
                sb.append(s);
                sb.append(",");
```

```
            }
            if(sb.length()>0){
                sb.deleteCharAt(sb.length()-1);
            }
        }
        return sb.toString();
    }
}
```

创建放置“开发技术调研”所用的页面 gather.jsp，并把选择的结果提交到 reg.jsp。关键代码如下：

```
<%@ page contentType="text/html;charset=UTF-8" language="java" %>
<html>
<head>
    <title>gather</title>
</head>
<body>
<form action="reg.jsp" method="post">
    <div>
        <h1>开发技术调研</h1>
        <hr/>
        <ul>
            <li>你经常用哪些编程语言开发程序</li>
            <li>
                <input type="checkbox" name="languages"        value="JAVA">JAVA
                <input type="checkbox" name="languages" value="PHP">PHP
                <input type="checkbox" name="languages" value=".NET">.NET
                <input type="checkbox" name="languages" value="Python">Python
            </li>
        </ul>
        <ul>
            <li>你目前掌握的技术:</li>
            <li>
                <input type="checkbox" name="technics" value="HTML">HTML
                <input type="checkbox" name="technics" value="JAVA BEAN">JAVA BEAN
                <input type="checkbox" name="technics" value="JSP">JSP
                <input type="checkbox" name="technics" value="Sping MVC">Spring MVC
            </li>
        </ul>
        <ul>
            <li>在学习中有哪一部分感觉有困难：</li>
            <li>
                <input type="checkbox" name="parts" value="JSP">JSP
                <input type="checkbox" name="parts" value="STRUTS">STRUTS
            </li>
        </ul>
        <input type="submit" value="提  交">
```

```
    </div>
</form>
</body>
</html>
```

创建名为 reg.jsp 的页面，用于对表单的技术调研结果进行处理，并将结果显示出来。关键代码如下：

```
<%@ page contentType="text/html;charset=UTF-8" language="java" %>
<html>
<head>
    <title>Reg</title>
</head>
<body>
<jsp:useBean id="paper" class="jspSamples.unit5.javaBeanSamples.Paper"></jsp:useBean>
<jsp:useBean id="convert" class="jspSamples.unit5.javaBeanSamples.Convert"></jsp:useBean>
<jsp:setProperty name="paper" property="*"></jsp:setProperty>
<div>
    <h1>开发技术调研结果</h1>
    <hr/>
    <ul>
        <li>
            你经常使用的编程语言： <%=convert.arr2Str(paper.getLanguages())%>
        </li>
        <li>
            你目前掌握的技术： <%=convert.arr2Str(paper.getTechnics())%>
        </li>
        <li>
            在学习中感觉有困难的部分： <%=convert.arr2Str(paper.getParts())%>
</div>
</li>
</ul>
</body>
</html>
```

reg.jsp 中实例化的两个 JavaBean 对象，均通过<jsp:useBean>标签进行实例化。运行 gather.jsp，对技术调研进行多选，单击“提交”按钮，请求发送到 reg.jsp 页面进行处理，结果如图 5-11 和图 5-12 所示。

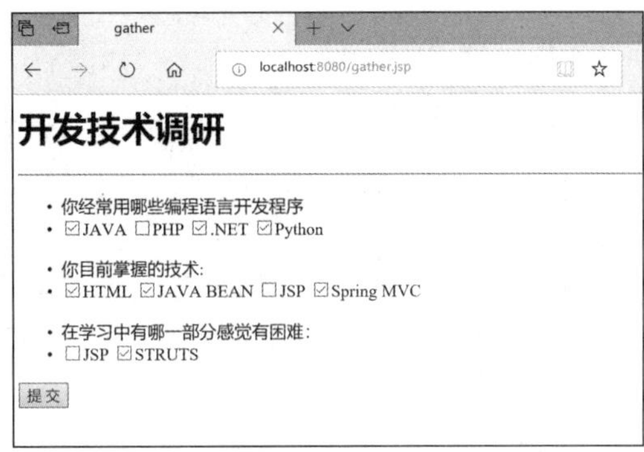

图 5-11

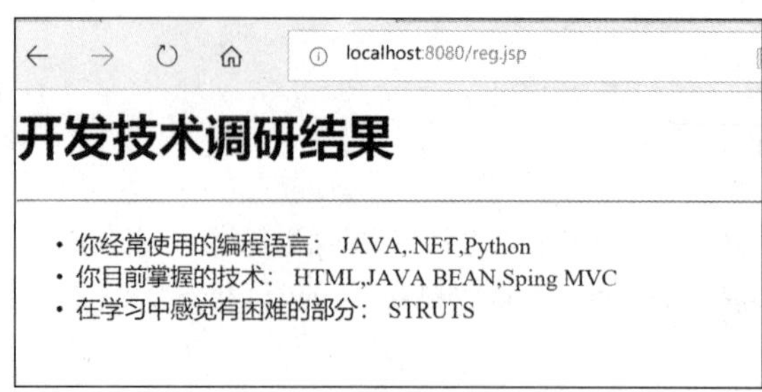

图 5-12

任务四　拓展实训

任务要求

本任务要求了解 JavaBean 的基本实现原理，通过拓展实训任务进一步熟悉 JavaBean 程序的编写和在 JSP 中的应用。

任务实现

初步认识了 IDEA 开发环境和 JavaBean 的实现原理之后，下面通过实训来练习 JavaBean 在 JSP 中的使用。

【实训 5-1】 加减乘除部分写在 JavaBean 中，在 JSP 页面中调用并显示计算结果。要求：按照 JavaBean 书写规则，实现加减乘除；主页面 index.jsp 中，设置供客户端输入的文本框，并显示最终计算结果，如图 5-13 所示。

图 5-13

参考代码——JavaBean 的实现 calculater.java：

```
package exp5;
public class calculater {
    float num1; // 参数1
    int operator; // 运算符
    float num2; // 参数2
    float result; // 运算结果
    public calculater() {
        super();
    }
```

```
    public float getNum1() {
        return num1;
    }
    public void setNum1(float num1) {
        this.num1 = num1;
    }
    public int getOperator() {
        return operator;
    }
    public void setOperator(int operator) {
        this.operator = operator;
    }
    public float getNum2() {
        return num2;
    }
    public void setNum2(float num2) {
        this.num2 = num2;
    }
    public float getResult() { // 计算式的运算结果
        float result1 = 0;
        try {
          switch (operator) {
              case 1:
                  result1 = num1 + num2;
                  break;
              case 2:
                  result1 = num1 - num2;
                  break;
              case 3:
                  result1 = num1 * num2;
                  break;
              case 4:
                  result1 = num1 / num2;
                  break;
              default:
                  break;
          }
        } catch (Exception e) {
          e.getMessage();
        }
        return result1;
    }
}
```

参考代码——JSP 的实现 index.jsp：

```
<%@ page language="java" contentType="text/html; charset=UTF-8"pageEncoding="UTF-8"%>
```

```
<!DOCTYPE html PUBLIC "-//W3C//DTD HTML 4.01 Transitional//EN" "http://www.w3.org/TR/html4/
loose.dtd">
<html>
<head>
  <meta http-equiv="Content-Type" content="text/html; charset=UTF-8">
  <title>Insert title here</title>
</head>
<body>
<jsp:useBean id="calculater" scope="request" class="exp5.calculater" />
<jsp:setProperty name="calculater" property="*" />
<form action="index.jsp" method="get">
  <hr align="left" style="width: 400px;">
  计算结果是：
  <span>
            <%
              if(request.getParameter("operator") != null){
               String operator = request.getParameter("operator");
               int oper = Integer.parseInt(operator);
               if(oper == 4 && calculater.getNum2() == 0){
                 out.print("出错，除数不能为零！");
               }else{
            %>
              <%=calculater.getNum1()%>
              <%
                if(oper == 1) out.print("+");
                if(oper == 2) out.print("-");
                if(oper == 3) out.print("*");
                if(oper == 4) out.print("/");
              %>
              <%=calculater.getNum2()%>
              =
              <%=calculater.getResult()%>
            <%
               }
              }
           %>
       </span>
  <!-- 表达式显示 -->
  <hr align="left" style="width: 400px;">
  <div align="left" style="width: 400px;">
    <p align="center">简单计算器</p>
    第一个参数：<input type="text" name="num1" /><br>
    <div style="padding-left: 100px;">
      <select name="operator">
        <option value="1">+</option>
```

```
            <option value="2">-</option>
            <option value="3">*</option>
            <option value="4">/</option>
        </select>
    </div>
    第二个参数：<input type="text" name="num2" /><br>
<input type="submit" value="计算" style="margin-left: 100px;" />
</div>
</form>
</body>
</html>
```

本章小结

本章主要介绍了 JavaBean 的种类及应用场景等，并且说明了 JavaBean 的实现原则与代码规则，通过获取 JavaBean 的属性信息、对 JavaBean 属性赋值、JavaBean 使用常见问题解决等若干个典型案例，重点详细讲解了 JavaBean 在各种技术场景中的实现与应用。在章节最后的拓展任务中，综合使用 JavaBean 技术实现页面计算器，使读者达到理解运用 JavaBean 技术实现综合业务的目标。

课后练习

1. 填空题

（1）__________和 JSP 相结合，可以实现表现层和商业逻辑层的分离。

（2）在 JSP 中可以使用__________操作来设置 Bean 的属性，也可以使用__________操作来获取 Bean 的值。

（3）JavaBean 有四个 scope，它们分别为__________、__________、__________和__________。

2. 选择题

（1）关于 JavaBean 正确的说法是（　　）。

A. Java 文件与 Bean 所定义的类名可以不同，但一定要注意区分字母的大小写

B. 在 JSP 文件中引用 Bean，其实就是用<jsp:useBean>语句

C. 被引用的 Bean 文件的文件名后缀为.java

D. Bean 文件放在任何目录下都可以被引用

（2）在 JSP 中调用 JavaBean 时不会用到的标记是（　　）。

A. <javabean>　　B. <jsp:useBean>

C. <jsp:setProperty>　　D. <jsp:getProperty>

（3）在项目中已经建立了一个 JavaBean，该类为 bean.Student，bean 具有 name 属性，则下面标签用法正确的是（　　）

A. <jsp:useBeanid="student" class="Student" scope="session"></jsp:useBean>

B. <jsp:useBeanid="student" class="Student" scope="session">hello student!</jsp:useBean>

C. <jsp:useBeanid="student" class="bean.Student" scope="session">hello student!</jsp:useBean>

D. <jsp:getProperty name="name" property="student"/>

（4）如果使用标记：<jsp:getProperty name="bean Name" property="property Name" />准备取出 bean 的属性的值，但 property Name 属性在 beanName 中不存在，也就是说在 beanName 中没有这样的属性名：property Name，也没有 getPropertyName()方法，那么会在浏览器中显示（　　）。

A. 错误页面　　B. null　　C. 0　　D. 什么也没有

（5）（　　）范围将使 Bean 一直保留到其到期或被删除为止。

A. page　　B. session　　C. application　　D. request

3. 判断题

（1）<jsp:getProperty>中的 name 及 property 区分大小写。（　　）

（2）在 JavaBean 中有很多方法，其中包含了主方法。（　　）

（3）JavaBean 中的属性既可以是 public 型的，也可以是 private 型的。（　　）

（4）编写 JavaBean 可以先不必加入到 JSP 程序中调用，而直接用 main 方法进行调试，调试好后就可以在 JSP 中使用了。（　　）

4. 简答题

（1）给已经实例化的 JavaBean 对象的属性赋值，有哪四种形式？

（2）JavaBean 的编写要求有哪些？

（3）JavaBean 的命名规范是什么？

5. 编程题

（1）编写一个封装学生信息的 JavaBean 对象，在 index.jsp 页面中调用该对象，并将学生信息输出到页面中。

（2）编写一个封装用户信息的 JavaBean 对象，通过操作 JavaBean 的动作标识，输出用户的注册信息。

（3）编写一个页面访问计数器的 JavaBean，在 index.jsp 页面中通过 JSP 动作标签实例化该对象，并将其放置于 application 范围中，实现访问计数器。

PART06

项目六

Servlet技术的应用

随着 Web 应用业务需求的增多，动态 Web 资源的开发变得越来越重要。目前，很多公司都提供了开发动态 Web 资源的相关技术，其中比较常见的有 ASP、PHP、JSP、Servlet 等。基于 Java 的动态 Web 资源开发，Sun 公司提供了 Servlet 技术，本章将针对 Servlet 技术的相关知识进行详细讲解。

课堂学习目标

- 了解 Servlet 的生命周期
- 掌握 Servlet 接口及其实现类的使用
- 掌握 Servlet 虚拟路径映射的配置

素养拓展

- 精益求精，一丝不苟

素养拓展

任务一 Servlet 技术概述

任务要求

本任务要求了解 Servlet 技术的基本知识，掌握其生命周期。

任务实现

（一）什么是 Servlet

Servlet 是 Server 与 Applet 的缩写，是服务器端小程序的意思，是 Sun 公司提供的一门用于开发动态 Web 资源的技术。Servlet 是基于 Java 语言的 Web 编程技术，部署在服务器端的 Web 容器里，获取客户端的访问请求，并根据请求生成响应信息返回到客户端。Java Web 应用程序请求处理流程如图 6-1 所示。 Servlet 是平台独立的 Java 类，编写一个 Servlet 实际上就是按照 Servlet 规范编写一个 Java 类。Servlet 被编译为平台独立的字节码，可以被动态地加载到支持 Java 技术的 Web 服务器中运行。

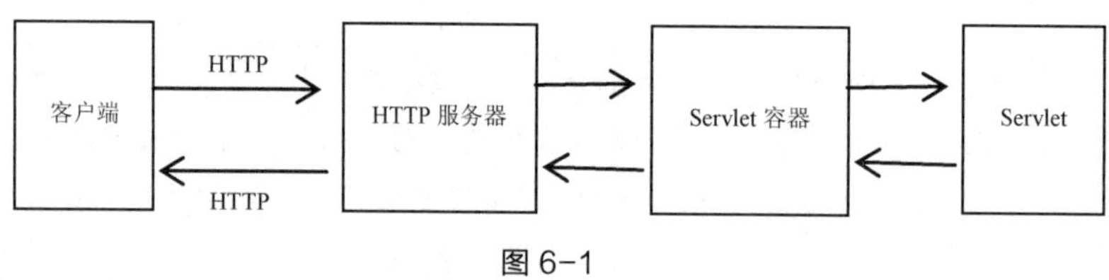

图 6-1

Servlet 容器（Servlet 引擎）是 Web 服务器或应用程序服务器的一部分，用于在发送的请求和响应之上提供网络服务，解码基于 MIME 的请求，格式化基于 MIME 的响应。

Servlet 不能独立运行，必须被部署到 Servlet 容器中，由容器来实例化和调用 Servlet 的方法，Servlet 容器在 Servlet 的生命周期内包容和管理 Servlet。

Servlet 的主要功能如下。

（1）读取客户端发送到服务器端的显式数据（表单数据）。

（2）读取客户端发送到服务器端的隐式数据（请求报头）。

（3）服务器端发送显式的数据到客户端（HTML）。

（4）服务器端发送隐式的数据到客户端（状态代码和响应报头）。

（二）Servlet 的生命周期

一个 Servlet 的生命周期由部署该 Servlet 的 Web 容器负责，除此之外，Web 容器还提供请求分发、安全、并发控制等服务。当特定的请求被容器映射到某个 Servlet 时，容器会做以下操作。

（1）若该 Servlet 的实例尚未被创建，那么 Web 容器就会加载该 Servlet 的.class 文件，然后创建该 class 的一个实例，并调用该实例的 init()方法来初始化该实例。

（2）调用初始化好的该实例的 Service()方法，传入 Request 和 Response 对象。

（3）最后如果不再需要该 Servlet，容器会调用该 Servlet 的 destroy()方法销毁该实例。

（4）Servlet 的生命周期从 Web 应用服务器开始运行时开始，以后会不断处理来自浏览器的访问请求，并将响应结果通过 Web 应用服务器返回给客户端，直到 Web 服务器停止运行，Servlet 才会被清除。

在 javax.servlet.Servlet 接口中，定义了针对 Servlet 生命周期最重要的三个方法，按照顺序依次是 init()、service()和 destroy()。

1. init()方法

该方法是 HttpServlet 类中的方法，可以在 Servlet 类中重写这个方法。init()方法的声明格式：

```
public void init(ServletConfig config) throws ServletException
```

Servlet 对象第一次被请求加载时，服务器创建一个 Servlet 对象，这个对象调用 init()方法完成必要的初始化工作。该方法在执行时，服务器会把一个 ServletConfig 类型的对象传递给 init()方法，这个对象被保存在 Servlet 对象中，直到 Servlet 对象被销毁。这个 ServletConfig 对象负责向 Servlet 传递服务设置信息，如果传递失败就会发生 ServletException，Servlet 对象就不能正常工作。init()方法只被调用一次，即在 Servlet 第一次被请求加载时调用该方法。

2. service()方法

该方法是 HttpServlet 类中的方法，可以在 Servlet 类中直接继承该方法或重写这个方法。service()方法的声明格式：

```
public void service (HttpServletRequest request,HttpServletResponse response) throw ServletException,
IOException
```

当 Servlet 对象成功创建和初始化之后，该对象就调用 service()方法来处理用户的请求并返回响应。服务器将两个参数传递给该方法：一个参数是 HttpServletRequest 类型的对象，该对象封装了用户的请求信息；另一个参数是 HttpServletResponse 类型的对象，用来响应用户的请求。和 init()方法不同，init()方法只被调用一次，而 service()方法可能被多次调用。我们已经知道，当后续的客户请求该 Servlet 对象服务时，服务器将启动一个新的线程，在该线程中，Servlet 对象调用 service()方法响应客户的请求，也就是说，每个客户的每次请求都导致 service()方法被调用执行，调用过程运行在不同的线程中，互不干扰。因此，不同线程的 service()方法中的局部变量互不干扰，一个线程改变了自己的 service()方法中局部变量的值，不会影响其他线程的 service()方法中的局部变量。

3. destroy()方法

当服务器关闭或 Web 应用被移除出容器时，Servlet 随着 Web 应用的销毁而销毁。在销毁 Servlet 之前，Servlet 容器会调用 Servlet 的 destroy()方法，以便让 Servlet 对象释放它所占用的资源。在 Servlet 的整个生命周期中，destroy()方法也只被调用一次。需要注意的是，Servlet 对象一旦创建就会驻留在内存中等待客户端的访问，直到服务器关闭，或 Web 应用被移除出容器时，Servlet 对象才会销毁。

Servlet 的生命周期如图 6-2 所示。

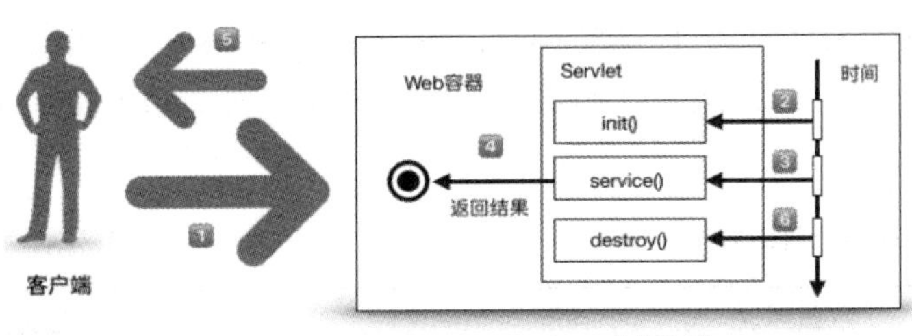

图 6-2

（三）Servlet 技术的特点

Servlet 程序在服务器端运行，动态地生成 Web 页面。与传统的 CGI 技术相比，Servlet 具有更高的效率，更容易使用，功能更强大，具有更好的可移植性，其主要特点包括以下几个方面。

（1）高效。在服务器上仅有一个 Java 虚拟机在运行，它的优势在于当多个来自客户端的请求进行访问时，Servlet 为每个请求分配一个线程而不是进程。

（2）方便。Servlet 提供了大量的实用工具例程，如处理很难完成的 HTML 表单数据，读取和设置 HTTP 头，处理 Cookie 和跟踪会话等。

（3）跨平台。Servlet 是用 Java 类编写的，可以在不同的操作系统平台和应用服务器平台下运行。

（4）功能强大。在 Servlet 中，许多使用传统 CGI 程序很难完成的任务都可以利用 Servlet 技术轻松完成。例如，Servlet 能够直接和 Web 服务器交互，而普通的 CGI 程序不能。Servlet 还能够在各个程序之间共享数据，使得数据库连接池之类的功能很容易实现。

（5）灵活性和可扩展性。采用 Servlet 开发的 Web 应用程序，由于 Java 类的继承性、构造函数等特点，使得其应用灵活，可随意扩展。

（6）共享数据。Servlet 之间通过共享数据可以很容易地实现数据库连接池。它能方便地实现管理用户请求、简化 Session 和获取前一页面信息的操作，而在 CGI 之间通信则很差。由于每个 CGI 程序的调用都开始一个新的进程，调用间的通信通常要通过文件进行，因而相当缓慢。同一台服务器上的不同 CGI 程序之间的通信也相当麻烦。

（7）安全。有些 CGI 版本有明显的安全弱点。即使是使用最新标准，系统也没有基本安全框架。而 Java 定义有完整的安全机制，包括 SSL/CA 认证、安全政策等规范。

任务二　编写 Servlet 类

任务要求

本任务要求掌握编写 Servlet 类的方法。

任务实现

（一）Servlet 类的结构

对于有一定 Java 基础的人，编写创建 Servlet 对象的类并不困难，因为编写一个创建 Servlet 对象的类就是编写一个特殊类的子类，这个特殊类就是 javax.servlet.http 包中的 HttpServlet 类。HttpServlet 类实现了 Servlet 接口，实现了响应用户的方法。HttpServlet 类的子类被习惯地称为一个 Servlet 类，这样的类创建的对象习惯地被称为一个 Servlet 对象。HttpServlet 的常用方法及其说明如表 6-1 所示。

表 6-1　HttpServlet 类的常用方法

方法声明	功能描述
protected void doGet(HttpServletRequest req, HttpServletResponse resp)	用于处理 GET 类型的 HTTP 请求的方法

续表

方法声明	功能描述
protected void doPost(HttpServletRequest req, HttpServletResponse resp)	用于处理 POST 类型的 HTTP 请求的方法

客户在使用表单提交数据到服务器的时候有两种方式可供选择，一种是 post，另一种是 get。可在<form>的 method 属性中指定提交的方式。如：<form action="inputForm" method="get">，如果不指定 method 属性，则会默认该属性为 get 方式。get 和 post 都能够提交数据，它们有什么不同呢?

（1）通过 get 方式提交的数据有大小的限制，通常在 1024B 左右。也就是说如果提交的数据很大，用 get 方法就需要小心；而 post 方式没有数据大小的限制，理论上传送多少数据都可以。

（2）通过 get 传递数据，实际上是将传递的数据按照"key，value"的方式跟在 URL 的后面来达到传送的目的；而 post 传递数据是通过 http 请求的附件进行的，将表单内各个字段与其内容放置在 HTML HEADER 内一起传送到 ACTION 属性所指的 URL 地址，在 URL 中并没有明文显示。

（3）通过 get 方式提交的数据安全性不高，而 post 方式更加安全。

（4）编码转换在 request 请求里面，get 方法得到的内容每一个都要进行编码转换，需要配置 Tomcat 编码设置，而 post 方法只要设置 request.setCharacterEncoding("UTF-8")就可以。

（5）对于 get 方式，服务器端用 Request.QueryString 获取变量的值；对于 post 方式，服务器端用 Request.Form 获取提交的数据。

在 Servlet 接口的 service（ServletRequest request，ServletResponse response）方法中有一个 ServletRequest 类型的参数。ServletRequest 类表示来自客户端的请求。当 Servlet 容器接收到客户端要求访问特定 Servlet 的请求时，容器先解析客户端的原始请求数据，把它包装成一个 ServletRequest 对象。当容器调用 Servlet 对象的 service()方法时，就可以把 ServletRequest 对象作为参数传给 service()方法。

ServletRequest 接口提供了一系列用于读取客户端请求数据的方法，如表 6-2 所示。

表 6-2　ServletRequest 接口的常用方法

方法声明	功能描述
getContentLength()	返回请求正文的长度。如果请求正文的长度未知，则返回-1
getContentType()	获得请求正文的 MIME 类型。如果请求正文的类型未知，则返回 null
getInputStream()	返回用于读取请求正文的输入流
getParameter(String name)	根据给定的请求参数名，返回来自客户请求中匹配的请求参数值
getReader()	返回用户读取字符串形式的请求正文的 Buffered Reader 对象
setAttribute(String name, Object object)	在请求范围内保存一个属性，参数 name 标识属性名，参数 object 标识属性值
getAttribute(String name)	根据 name 参数给定的属性名，返回请求范围内匹配的属性值
removeAttribute(String name)	从请求范围内删除一个属性

HttpServletRequest 接口是 ServletRequest 接口的子接口。HttpServlet 类的重载 service()方法及 doGet()和 doPost()等方法都有一个 HttpServletRequest 类型的参数，如：

```
protected void service(HttpServletRequest req, HttpServletResponse resp)
throws ServletException, IOException{........}
```

HttpServletRequest 接口提供了用于读取 HTTP 请求中相关信息的方法，如表 6-3 所示。

表 6-3　HttpServletRequest 接口的常用方法

方法声明	功能描述
getContextPath()	返回客户端所请求访问的 web 应用的 URL 入口。例如，如果客户端访问的 URL 为 http://localhost:8080/helloapp/info，那么该访问返回 "/helloapp"
getCookies()	返回 HTTP 请求中的所有的 Cookie
getHeader（String name）	返回 HTTP 请求头部的特定项
getHeaderNames()	返回一个 Enumeration 对象，它包含 HTTP 请求头部的所有项目名
getMethod()	返回 HTTP 请求方式，如 post 或 get
getRequestURI()	返回 HTTP 请求的头部的第 1 行的 URI
getQueryString()	返回 HTTP 请求中的查询字符串，即 URL 中的"？"后面的内容。例如，如果客户端访问的 URL 为 http://localhost:8080/helloapp/info?username=tom，那么该访问返回 "username=tom"

根据 SUN 的 Servlet API 来创建的 Servlet，无须费力地解析原始 HTTP 请求。解析原始 HTTP 请求的工作完全由 Servlet 容器来代劳。Servlet 容器把 HTTP 请求包装成 HttpServletRequest 对象，Servlet 只需调用该对象的 getXXX()方法，就能轻轻松松地读取到 HTTP 请求中的各种数据。

在 Servlet 接口的 service(ServletReuqest req,ServletResponse res)方法中还有一个 ServletResponse 类型的参数。Servlet 通过 ServletResponse 对象来生成响应结果。当 Servlet 容器接收到客户端要求访问特定 Servlet 的请求时，容器会创建一个 ServletResponse 对象，并把它作为参数传给 Servlet 的 service()方法。

在 ServletResponse 接口中定义了一系列与生成响应相关的方法，如表 6-4 所示。

表 6-4　ServletResponse 接口的常用方法

方法声明	功能描述
setCharacterEncoding(String charset)	设置响应正文的字符编码。响应正文的默认字符编码为 ISO-8859-1
setContentLength(int len)	设置响应正文的长度
setContentType（String type）	设置响应正文的 MIME 类型
getCharacterEncoding()	返回响应正文的字符编码
getContentType()	返回响应正文的 MIME 类型
setBufferSize(int size)	设置用于存放响应正文数据的缓存区的大小
getBufferSize()	获得用于存放正文数据的缓存区的大小

续表

方法声明	功能描述
reset()	清空缓存区内的正文数据，并且清空响应状态代码及响应头
resetBuffer()	仅仅清空缓存区内的正文数据，不清空响应状态代码及响应头
flushBuffer()	强制性地把缓存区内的响应正文数据发送到客户端
isCommitted()	返回一个 boolean 类型的值。如果为 true，表示缓存区内的数据已经提交给客户，即数据已经发送到客户端
getOutputStream()	返回一个 ServletOutputStream 对象，Servlet 用它来输出二进制的正文数据
getWriter()	返回一个 PrintWriter 对象，Servlet 用它来输出字符串形式的正文数据

Servlet 通过 ServletResponse 对象主要产生 HTTP 响应结果的正文部分。

为了提高输出数据的效率，ServletOutputStream 和 PrintWriter 先把数据写到缓存区内。当缓存区内的数据被提交到客户后，ServletResponse 的 isCommitted()方法返回 true。

在以下几种情况下，缓存区内的数据会被提交给客户，即数据被发送到客户端。

（1）当缓存区内的数据已满时，ServletOutputStream 或 PrintWriter 会自动把缓存区内的数据发送给客户端，并且清空缓存区。

（2）Servlet 调用 ServletResponse 对象的 flushBuffer()方法。

（3）Servlet 调用 ServletOutputStream 对象或 PrintWriter 对象的 flush()或 close()方法。

为了确保 ServletOutputStream 或 PrintWriter 输出的所有数据都会被提供给客户，比较安全的做法是在所有的数据都输出完毕后，调用 ServletOutputStream 或 PrintWriter 的 close()方法。

需要注意的是，如果要设置响应正文的 MIME 类型和字符编码，必须先调用 ServletResponse 对象的 setContentType()和 setCharacterEncoding()方法，然后再调用 ServletResponse 的 getOutputStream()或 getWriter()方法，或者提交缓存区内的正文数据。只有满足这样的操作顺序，所做的设置才能生效。

HttpServletResponse 接口是 ServletResponse 的子接口，HttpServlet 类的重载 service()方法及 doGet()和 doPost()方法都有一个 HttpServletResponse 类型的参数：

```
protected void service(HttpServletRequest req,HttpServletResponse resp)
throws ServletException, IOException{........}
```

HttpServletResponse 接口提供了与 HTTP 协议相关的一些方法，Servlet 可通过这些方法来设置 HTTP 响应头或向客户端写 Cookie。

HttpServletResponse 接口的常用方法如表 6-5 所示。

表 6-5　HttpServletResponse 接口的常用方法

方法声明	功能描述
addHeader(String name，String value)	向 HTTP 响应头中加入一项内容
sendError(int sc)	向客户端发送一个代表特定错误的 HTTP 响应状态码

续表

方法声明	功能描述
sendError(int sc, String msg)	向客户端发送一个代表特定错误的 HTTP 响应状态码，并且发送具体的错误信息
setHeader(String name, String value)	设置 HTTP 响应头中的一项内容。如果在响应头中已经存在这项内容，那么原先所做的设置将被覆盖
setStatus(int sc)	设置 HTTP 响应的状态代码
addCookie(Cookie cookie)	向 HTTP 响应中加入一个 Cookie

需要注意的是：ServletResponse 中响应正文的默认 MIME 类型为 text/plain，即纯文本类型。而 HttpServletResponse 中响应正文的默认 MIME 类型为 text/html，即 HTML 文档类型。

（二）建立 Servlet 类

【例 6-1】 编写一个简单的 Servlet 类。

1. 创建 Java Web 工程

本书编写 Servlet 的开发工具为 Intellij IDEA，在工具的主界面中选择 File -->New -->Project，然后在 Project Name 文本框中输入自己定制的工程名称，创建一个 Servlet 工程，如图 6-3 所示，单击 Finish 按钮后出现编辑页面。其中 src 是放置源代码包的，下面存放工程源代码，默认新建一个 index.jsp 页面，web.xml 位于 WEB-INF 文件夹下，用于存放 Servlet 的配置信息。

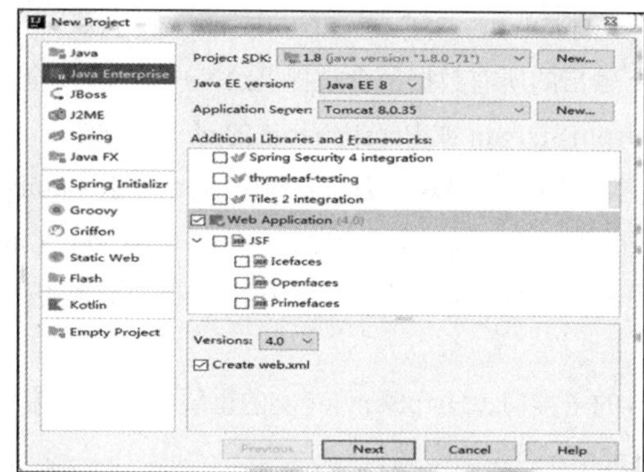

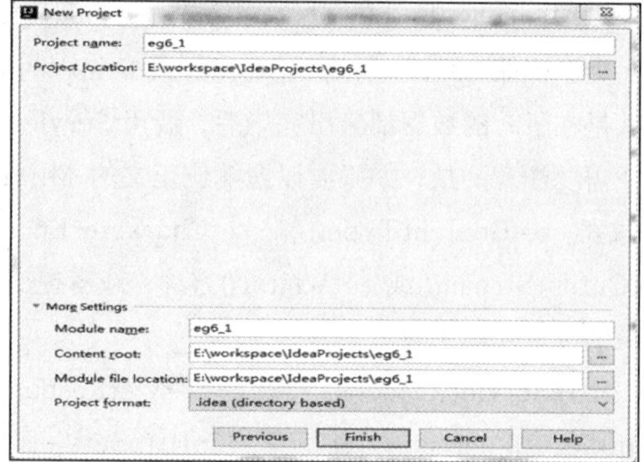

图 6-3

2. Web 工程设置

（1）在 WEB-INF 目录下单击右键，选择 New --> Directory，创建 classes 和 lib 两个目录，其中 classes 目录用于存放编译后的 class 文件，lib 目录用于存放依赖的 jar 包，如图 6-4 所示。

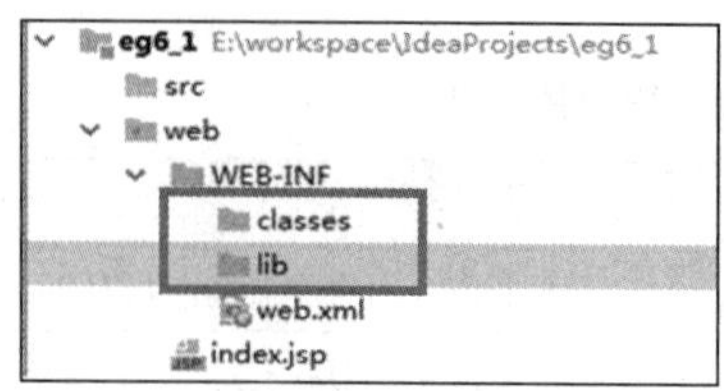

图 6-4

（2）在工具的主界面中选择 File --> Project Structure...，进入 Project Structure 对话框，单击 Modules -->选中项目“JavaWeb” --> 切换到 Paths 选项卡 -->勾选“Use module compile output path”，将 Output path 和 Test output path 都改为之前创建的 classes 目录，即将后面编译的 class 文件默认生成到 classes 目录下，如图 6-5 所示。

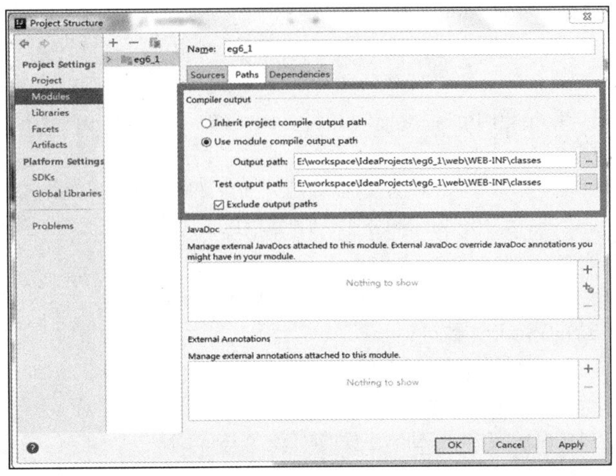

图 6-5

（3）单击 Modules --> 选中项目“eg6_1” --> 切换到 Dependencies 选项卡 --> 单击右边的“+”，选择 “JARs or directories...”，选择创建的 lib 目录，选择 Jar Directory。

（4）配置打包方式 Artifacts：单击 Artifacts 选项卡，IDEA 会为该项目自动创建一个名为“eg6_1:war exploded”的打包方式，表示打包成 war 包，并且是文件展开式的，输出路径为当前项目下的 out 文件夹，保持默认即可。

3. 建立 Servlet 类

右击 src 目录 --> 选中 New-->Servlet，新建一个 servlet，命名为 FirstServlet，如图 6-6 所示。

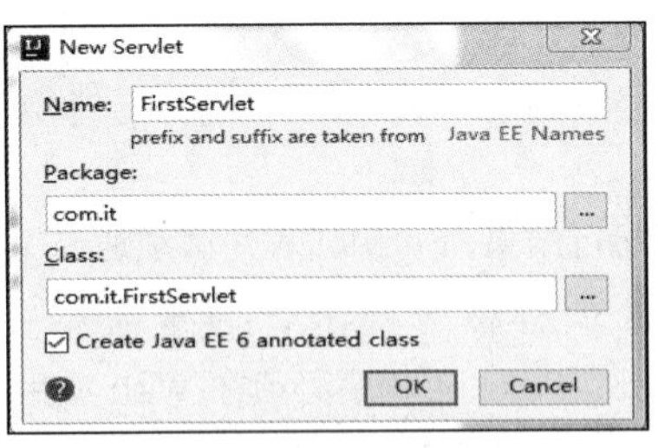

图 6-6

打开 FirstServlet.java 程序，写入如下代码，其中重写 doPost()和 doGet()方法中的一个；定义初始化的 init()方法，此方法用于获取资源文件里面的初始化信息；定义清除资源的 destroy()方法。

```
public class FirstServlet extends HttpServlet {
    public FirstServlet() {
        super();
    }
```

```
    @Override
    public void destroy() {
        super.destroy();
    }
    @Override
    public void init() throws ServletException {
        super.init();
    }
        protected void doPost(HttpServletRequest request, HttpServletResponse response) throws ServletException, IOException {
            response.setCharacterEncoding("UTF-8");      //设置响应的字符集格式为UTF-8
            response.setContentType("text/html");  //设置响应正文的MIME类型
            PrintWriter out = response.getWriter();
    //返回一个PrintWriter对象，Servlet使用它来输出字符串形式的正文数据
            //以下为输出的HTML正文数据
            out.println("<!DOCTYPE HTML PUBLIC \"-//W3C//DTD HTML 4.01 Transitional//EN\">");
            out.println("<HTML>");
            out.println("<HEAD><TITLE>动态生成的HTML文档</TITLE></HEAD>");
            out.println("<BODY>");
            out.println("<table border='0' align='center'>");
            out.println("<tr><td bgcolor='skyblue'colspan=2>动态生成HTML文档</td></tr>");
            out.println("</table>");
            out.println("</BODY>");
            out.println("</HTML>");
            out.flush();
            out.close();
        }
        protected void doGet(HttpServletRequest request, HttpServletResponse response) throws ServletException, IOException {
            doPost(request, response);
        }
    }
```

4. 注册和运行 Servlet

如果要用浏览器打开并查看运行结果，Servlet 程序必须通过 Web 服务器和 Servlet 容器来启动运行。Servlet 程序的存储目录有特殊要求，通常需要存储在<Web 应用程序目录>\WEB-INF\classes 目录中。另外，Servlet 程序必须在 Web 应用程序的 web.xml 文件中进行注册和映射其访问路径，才可以被 Servlet 容器加载和被外界访问。

（1）注册和映射 Servlet。在 web.xml 文件中，<servlet>元素用于注册 Servlet，<servlet>元素中包含两个主要的子元素，即<servlet-name>和<servlet-class>，分别用于设置 Servlet 的注册名称和指定 Servlet 的完整类名。

<servlet-mapping>元素用于映射已经注册的 Servlet 的对外访问路径，客户端将使用映射路径访问 Servlet。<servlet-mapping>元素中含有两个子元素，即<servlet-name>和<url-pattern>，分别用于指定 Servlet 的注册名称和设置 Servlet 的访问路径。

web.xml 的相关代码如下所示：

```
<servlet>
<!-- 声明Servlet对象 -->
        <servlet-name>FirstServlet</servlet-name>
        <!-- 上面一句指定Servlet对象的名称 -->
        <servlet-class>com.it.FirstServlet</servlet-class>
        <!-- 上面一句指定Servlet对象的完整位置，包含包名和类名 -->
    </servlet>
<servlet-mapping>
<!-- 映射Servlet -->
        <servlet-name>FirstServlet</servlet-name>
        <!--<servlet-name>与上面<Servlet>标签的<servlet-name>元素相对应，不可以随便起名 -->
        <url-pattern>/firstServlet</url-pattern>
        <!-- 上面一句话用于映射访问URL -->
</servlet-mapping>
```

（2）运行 Servlet。启动 Tomcat，在浏览器中输入 http://localhost:8080/firstServlet，显示如图 6-7 所示的效果。

图 6-7

任务三　编写 Web.xml 配置文件

任务要求

本任务要求掌握 Servlet 的多种配置方法，以及掌握读取配置文件的过程。

任务实现

（一）配置虚拟路径

在 web.xml 文件中，一个<servlet-mapping>元素用于映射一个 Servlet 的对外访问路径，该路径也称为虚拟路径。例如，<url-pattern>/firstServlet</url-pattern>，其中"/firstServlet"就是一个虚拟路径。创建的 Servlet 只有在 web.xml 中映射了虚拟路径，客户端才能访问。但是，在映射 Servlet 时有一些细节问题需要注意，比如 Servlet 的多重映射，在映射路径中使用通配符等。下面针对这些问题进行详细解释，具体如下。

1. Servlet 的多重映射

Servlet 的多重映射是指同一个 Servlet 可以被映射成多个虚拟路径，即客户端可以通过多个路径访问同一个 Servlet，具体有如下两种方法：①可以配置多个<servlet-mapping>标签；②可以在<servlet-mapping>标签中配置多个<url-pattern>标签。

```
<!-- 第一种方法-->
<servlet>
    <servlet-name>FirstServlet</servlet-name>
    <servlet-class>com.it.FirstServlet</servlet-class>
</servlet>
<servlet-mapping>
    <servlet-name>FirstServlet</servlet-name>
    <url-pattern>/firstServlet01</url-pattern>
</servlet-mapping>
<servlet-mapping>
    <servlet-name>FirstServlet</servlet-name>
    <url-pattern>/firstServlet02</url-pattern>
</servlet-mapping>

<!-- 第二种方法-->
<servlet>
    <servlet-name>SecondServlet</servlet-name>
    <servlet-class>com.it.SecondServlet</servlet-class>
</servlet>
<servlet-mapping>
    <servlet-name>SecondServlet</servlet-name>
    <url-pattern>/secondServlet01</url-pattern>
    <url-pattern>/secondServlet02</url-pattern>
</servlet-mapping>
```

2. 映射路径中使用通配符

有时候希望某个目录下的所有路径都可以访问同一个 Servlet，这时，可以在 Servlet 映射的路径中使用通配符“*”，如下所示：

```
<url-pattern>/servlet/*</url-pattern>
```

这种属于路径匹配，通配符“*”为后缀，/servlet/firstServlet01、/servlet/firstServlet02 都与/servlet/*匹配。

```
<url-pattern>*.do</url-pattern>
```

这种属于扩展名匹配，通配符“*”为前缀，/servlet/firstServlet.do、/firstServlet.do 都与*.do 匹配。

```
<url-pattern>/*</url-pattern>
```

这种属于完全匹配，通配符“*”为后缀，匹配所有路径。

需要注意的是：通配符要么在开头，要么在结尾，不能在中间，<url-pattern>/*.do</url-pattern>是错误的。

如果不使用通配符，那么 url 标签的内容必须以"/"开头，<url-pattern>firstServlet </url-pattern>是错误的。

【例 6-2】 三种通配符匹配方式的优先级。

```
<servlet>
    <servlet-name>FirstServlet</servlet-name>
    <servlet-class>com.it.FirstServlet</servlet-class>
</servlet>
<servlet-mapping>
    <servlet-name>FirstServlet</servlet-name>
    <url-pattern>*.do</url-pattern>
</servlet-mapping>

<servlet>
    <servlet-name>SecondServlet</servlet-name>
    <servlet-class>com.it.SecondServlet</servlet-class>
</servlet>
<servlet-mapping>
    <servlet-name>SecondServlet</servlet-name>
    <url-pattern>/*</url-pattern>
</servlet-mapping>
```

根据配置，当在浏览器端访问 http://localhost:8080/hello.do 时，FirstServlet 和 SecondServlet 都能够匹配。根据匹配的范围越大，优先级越低的原则，SecondServlet 匹配得更加准确，范围更小，所以访问的是 SecondServlet 这个 Servlet。

（二）ServletConfig 和 ServletContext

1. ServletConfig

在 Servlet 的配置文件中，可以使用一个或多个<init-pararn>标签为 servlet 配置一些初始化参数。在 Web 容器初始化 Servlet 实例时，都会为这个 Servlet 准备一个唯一的 ServletConfig 实例，Web 容器会从部署的描述文件中"读出"该 Servlet 类的初始化参数，并设置到 ServletConfig 实例中，然后再把这个 ServletConfig 实例传递给该 Servlet 实例的 init()方法。进而就可以通过 ServletConfig 对象得到当前 servlet 的初始化参数信息。

首先，需要创建私有变量：

```
private ServletConfig config = null;
```

其次，要重写 init()方法，传入 config，令 this.config = config;，从而获得 ServletConfig 对象。

最后，可以通过 ServletConfig 接口中提供的 getInitParameter(String name)方法来获取指定名称的初始化参数的字符串值。

【例 6-3】 利用 ServletConfig 读取配置信息的内容。

web.xml 的代码如下：

```
<servlet>
    <servlet-name>ServletConfigTest</servlet-name>
```

```
        <servlet-class>com.it.ServletConfigTest</servlet-class>
        <init-param>
            <param-name>charset</param-name>
            <param-value>utf-8</param-value>
        </init-param>
</servlet>
<servlet-mapping>
        <servlet-name>ServletConfigTest</servlet-name>
        <url-pattern>*.do</url-pattern>
</servlet-mapping>
```

ServletConfigTest.java 文件的代码如下：

```
public class ServletConfigTest extends HttpServlet {
    private  ServletConfig config;
    public ServletConfigTest() {
        super();
    }
    @Override
    public void destroy() {
        super.destroy();
    }
    @Override
    public void init(ServletConfig config) throws ServletException {
        super.init(config);
        this.config=config;
    }
    protected void doPost(HttpServletRequest request, HttpServletResponse response) throws
ServletException, IOException {

        String charset=  this.config.getInitParameter("charset");
        response.setCharacterEncoding(charset);     //设置响应的字符集格式为UTF-8
        response.setContentType("text/html");  //设置响应正文的MIME类型
        PrintWriter out = response.getWriter();
//返回一个PrintWriter对象，Servlet使用它来输出字符串形式的正文数据
        //以下为输出的HTML正文数据
        out.println("<HTML>");
        out.println("  <HEAD><TITLE>ServletConfig </TITLE></HEAD>");
        out.println("  <BODY>");
        out.println("获取InitParamServlet的初始化参数\"encoding\"的字符串值："+charset);
        out.println("  </BODY>");
        out.println("</HTML>");
        out.flush();
        out.close();
    }
    protected void doGet(HttpServletRequest request, HttpServletResponse response) throws
ServletException, IOException {
```

```
doPost(request,response);
    }
}
```

运行效果如图 6-8 所示。

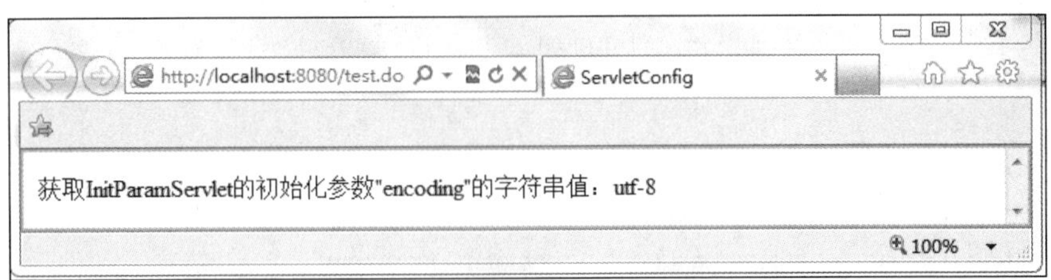

图 6-8

Servlet 的初始化参数只是针对当前这个 Servlet 类有效，在本 Servlet 类中只能获取自身的初始化参数，无法获取其他 Servlet 类的初始化参数。

2. ServletContext

如果在多个 Servlet 类要获取相同的初始化参数值，给每个 Servlet 都配置相同的初始化参数值显然是不太可取的。这时就可以把参数配置成 Web 应用上下文初始化参数。

Web 容器部署某个 Web 应用程序后，会为每个 Web 应用程序创建一个 ServletContext 实例。通过这个 ServletContext 实例就可以获取到所有的 Web 应用上下文初始化参数的值。

ServletContext 可以被认为是对于 Web 应用程序的一个整体性存储区域。每一个 Web 应用程序都只有一个 ServletContext 实例，存储在 ServletContext 之中的对象将一直被保留，直到它被删除。

ServletContext 对象可以通过 ServletConfig.getServletContext()方法获得对 ServletContext 对象的引用，也可以通过 this.getServletContext()方法获得其对象的引用。

【例 6-4】 利用 ServletContext 读取配置信息的内容。

web.xml 的代码如下：

```
<context-param>
    <param-name>url</param-name>
    <param-value>jdbc:mysql:/localhost/dbName</param-value>
</context-param>
<context-param>
    <param-name>username</param-name>
    <param-value>root</param-value>
</context-param>
<context-param>
    <param-name>password</param-name>
    <param-value>123456</param-value>
</context-param>
```

ServletContextTest1.java 文件的代码如下：

```
public class ServletContextTest1 extends HttpServlet {
    private ServletConfig config;
    private ServletContext context;
```

```
        @Override
        public void init(ServletConfig config) throws ServletException {
            super.init(config);
            this.config = config;
        }
        protected void doPost(HttpServletRequest request, HttpServletResponse response) throws
ServletException, IOException {

            context = this.config.getServletContext();
            String url = this.context.getInitParameter("url");
            String username = this.context.getInitParameter("username");
            String password = this.context.getInitParameter("password");
            response.setCharacterEncoding("UTF-8");      //设置响应的字符集格式为UTF-8
            response.setContentType("text/html");  //设置响应正文的MIME类型
            PrintWriter out = response.getWriter();
    //返回一个PrintWriter对象，Servlet使用它来输出字符串形式的正文数据
            //以下为输出的HTML正文数据
            out.println("<HTML>");
            out.println("  <HEAD><TITLE>ServletContext </TITLE></HEAD>");
            out.println("  <BODY>");
            out.println("获取连接数据库的初始化参数 ："+"<br/>");
            out.println("url:" + url+"<br/>");
            out.println("username:" + username+"<br/>");
            out.println("password:" + password+"<br/>");
            out.println("  </BODY>");
            out.println("</HTML>");
            out.flush();
            out.close();
        }
        protected void doGet(HttpServletRequest request, HttpServletResponse response) throws
ServletException, IOException {
            doPost(request, response);
        }
    }
```

运行效果如图 6-9 所示。

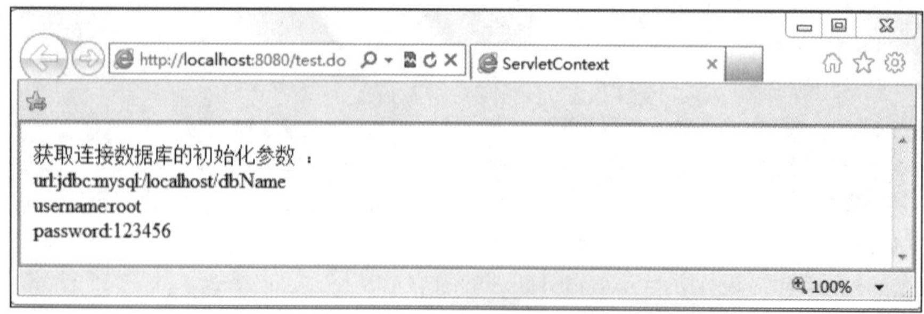

图 6-9

由于一个 Web 应用中的所有 Servlet 共享同一个 ServletContext 对象，因此 Servlet 对象之间可以通过 ServletContext 对象来实现通信。在 serlvet 中，可以使用如下语句来设置数据共享：

```
ServletContext context =this.getServletContext(); //servletContext域对象
context.setAttribute("data","共享数据"); //向域中存了一个data属性
```

在另一个 servlet 中，可以使用如下语句来获取域中的 data 属性：

```
ServletContext context =this.getServletContext();
String value = (String)context.getAttribute("data");    //获取域中的data属性
System.out.println(value);
```

任务四　Servlet 类的访问

任务要求

本任务要求掌握 Servlet 类的访问方法。

任务实现

（一）通过表单访问 Servlet 类

在前面已经学习过 get 和 post 方法的区别，以及应用 request 对象获取 HTML 表单数据的方法。Servlet 同样可以自动完成对 HTML 表单数据的读取操作。在 Servlet 中只需要简单地调用 HttpServletRequest 的 getParameter 方法，在调用参数中提供表单变量的名字即可。如果指定的表单变量存在，但没有值，getPammeter 返回空字符串；如果指定的表单变量不存在，则返回 null。如果表单变量可能对应多个值，可以用 getParameterValues 来取代 getParameter。getParameterValues 能够返回一个字符串数组。

微课：通过表单访问 Servlet

另外，在调试环境中，需要获得完整的表单变量名字列表。利用 HttpServletRequest 的 getParameterNames 方法可以方便地实现这一点。getParameterNames 返回的是一个 Enumeration，其中的每一项都可以转换为调用 getParameter 的字符串。

【例 6-5】 利用 Servlet 完成读取注册表单中的信息。

index.jsp 页面的代码如下：

```
<form action="RegisterServlet" method="post">
    <table width="80%" border="0" align="center" bgcolor="#0099ff">
        <tr>
            <th colspan="2" scope="col">用户注册</th>
        </tr>
        <tr bgcolor="#FFFFFF">
            <th scope="row">用户名：</th>
            <td><input name="userName" type="text" id="userName"></td>
        </tr>
        <tr bgcolor="#FFFFFF">
            <th scope="row">密码：</th>
```

```
            <td><input name="userPwd" type="password" id="userPwd"></td>
        </tr>
        <tr bgcolor="#FFFFFF">
            <th scope="row">电子邮箱：</th>
            <td><input name="email" type="text" id="email"></td>
        </tr>
        <tr bgcolor="#FFFFFF">
            <th scope="row">性别：</th>
            <td><input type="radio" name="userSex" id="userMale" value="男">男
                <input type="radio" name="userSex" id="userFemale" value="女">女
            </td>
        </tr>
        <tr bgcolor="#FFFFFF">
            <th scope="row">教育程序：</th>
            <td>
                <select name="userEducation" id="userEducation" >
                    <option value="研究生">研究生</option>
                    <option value="本科">本科</option>
                    <option value="专科">专科</option>
                    <option value="高中">高中</option>
                </select>
            </td>
        </tr>
        <tr bgcolor="#FFFFFF">
            <th scope="row"> </th>
            <td><input type="submit" name="Snbmit" value="提交">
                <input type="reset" name="Reset" value="重置">
            </td>
        </tr>
    </table>
</form>
```

RegisterServlet.java 页面的代码如下：

```
public class RegisterServlet extends HttpServlet {
    protected void doPost(HttpServletRequest request, HttpServletResponse response) throws
ServletException, IOException {
        response.setContentType("text/html");
        response.setCharacterEncoding("utf-8");
        request.setCharacterEncoding("utf-8");
        String userName = request.getParameter("userName");
        String userPwd = request.getParameter("userPwd");
        String email = request.getParameter("email");
        String userSex = request.getParameter("userSex");
        String userEducation = request.getParameter("userEducation");
        PrintWriter out = response.getWriter();
        out.println("<HTML>");
```

```
            out.println("    <HEAD><TITLE>用户注册结果 </TITLE></HEAD>");
            out.println("    <BODY><br/>");
            out.print("<h3>用户注册结果</h3>");
            out.println("<table border=1 align=left");
            out.println("<tr><th>参数</th><th>参数值</th></tr>");
            out.println("<tr><td>userName</td><td>" + userName + "</td></tr>");
            out.println("<tr><td>userPwd</td><td>" + userPwd + "</td></tr>");
            out.println("<tr><td>email</td><td>" + email + "</td></tr>");
            out.println("<tr><td>userSex</td><td>" + userSex + "</td></tr>");
            out.println("<tr><td>userEducation</td><td>" + userEducation + "</td></tr>");
            out.println("</table></BODY>");
            out.println("</HTML>");
            out.flush();
            out.close();   ;
        }
        protected void doGet(HttpServletRequest request, HttpServletResponse response) throws ServletException, IOException {
             doPost(request, response);
        }
    }
    web.xml
    <servlet>
            <servlet-name>RegisterServlet</servlet-name>
            <servlet-class>com.it.RegisterServlet</servlet-class>
        </servlet>
        <servlet-mapping>
            <servlet-name>RegisterServlet</servlet-name>
            <url-pattern>/RegisterServlet</url-pattern>
        </servlet-mapping>
```

运行效果如图 6-10 所示。

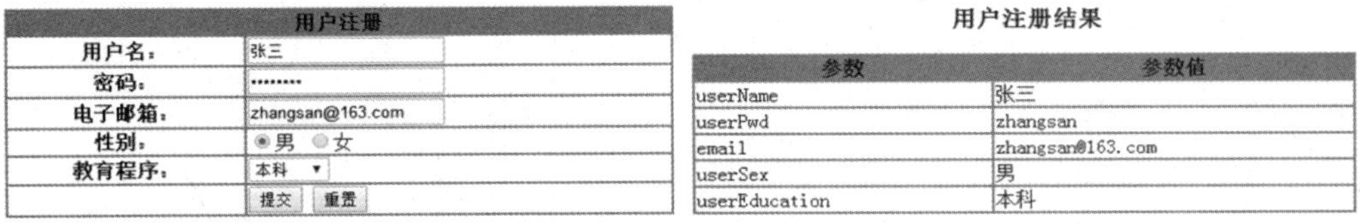

用户注册	
用户名：	张三
密码：	••••••••
电子邮箱：	zhangsan@163.com
性别：	◉男 ○女
教育程序：	本科 ▾
	提交 重置

用户注册结果

参数	参数值
userName	张三
userPwd	zhangsan
email	zhangsan@163.com
userSex	男
userEducation	本科

图 6-10

（二）通过 JSP 页面访问 Servlet 类

尽管可以用在浏览器的地址栏中直接键入 servlet 对象的请求格式来运行一个 servlet，但可能会经常通过一个 JSP 页面来请求一个 servlet。也就是说，可以让 JSP 页面负责数据的显示，而让 servlet 去做与处理数据有关的事情。

【例 6-6】 单击超链接，利用 Servlet 对象输出英文字母表。

index.jsp 页面的代码如下：

```
<html>
  <head>
    <title>英文字母表</title>
  </head>
  <body>
<font size="3">
  单击超链接查看英文字母表:<br>
  <a href="ShowLetterServlet">查看英文字母表</a>
</font>
  </body>
</html>
```

ShowLetterServlet.java 页面的代码如下：

```
public class ShowLetterServlet extends HttpServlet {
    protected void doPost(HttpServletRequest request, HttpServletResponse response) throws Servlet
Exception, IOException {
        response.setContentType("text/html;charset=UTF-8");
        PrintWriter out = response.getWriter();
        out.println("<html><body>");
        out.println("<br>小写字母:");
        for (char ch = 'a'; ch <= 'z'; ch++) {
          out.printf("%-3c", ch);
        }
        out.println("<br>大写字母:");
        for (char ch = 'A'; ch <= 'Z'; ch++) {
          out.printf("%3c", ch);
        }
        out.println("</body></hemtl>");
    }
    protected void doGet(HttpServletRequest request, HttpServletResponse response) throws
ServletException, IOException {
        doPost(request, response);
    }
}
web.xml
    <welcome-file-list>
        <welcome-file>index.jsp</welcome-file>
    </welcome-file-list>
    <servlet>
        <servlet-name>ShowLetterServlet</servlet-name>
        <servlet-class>com.it.ShowLetterServlet</servlet-class>
    </servlet>
    <servlet-mapping>
        <servlet-name>ShowLetterServlet</servlet-name>
        <url-pattern>/ShowLetterServlet</url-pattern>
    </servlet-mapping>
```

运行效果如图 6-11 所示。

图 6-11

Servlet 体系是基于 B/S 架构开发 web 应用程序，使用 Servlet 类将 HTTP 请求和响应封装在标准 Java 类中来实现各种 web 应用方案。当大量的 B/S 架构程序开发出来以后出现了很多问题：首先 Servlet 类有大量冗余代码，其次是开发 Servlet 没法做到有精美的页面效果。所以 SUN 公司提出将服务端代码添加在已经设计好的静态页面上，经过 JSP 容器对 JSP 文件进行自动解析并转换成 Servlet 类来交给 web 服务器运行。

所以 JSP 在本质上就是 Servlet，但是两者的创建方式不一样。Servlet 都是由 Java 程序代码构成，用于流程控制和事务处理，通过 Servlet 来生成动态网页很不直观。而 JSP 由 HTML 代码和 JSP 标签构成，可以方便地编写动态网页。在 struts 框架中，JSP 位于 MVC 设计模式的视图层，而 Servlet 位于控制层。

任务五　拓展实训

任务要求

在实际开发中，希望读取 web 应用中的一些资源文件，比如配置文件、图片等。为此，在 ServletContext 接口中提供了一些读取 web 资源的方法，这些方法是依靠 Servlet 容器来实现的。

任务实现

在项目中创建一个.properties 的配置文件，其主要作用是通过修改配置文件可以方便地修改代码中的参数，实现不用改 class 文件即可灵活地变更参数。Servlet 容器根据资源文件相对于 web 应用的路径，返回关联资源文件的 IO 流、资源文件在文件系统的绝对路径等。

ServletContext 读取的.properties 文件一般放在如下位置：①直接放在 Web 下面；②放在 Web 下面的某个文件夹下面；③放在 WEB-INF 下面的某个文件夹下面。

特别注意，不要把.properties 文件直接放在 WEB-INF 下面（因为这样取得的结果是 null）。

下面来讲解 ServletContext 读取 web 应用中的资源文件的方法。

1. 获取真实路径

ServletContext 接口的 getRealPath（String path）方法返回的是资源文件在服务器文件系统上的真实路径（带有盘符）。参数 path 代表资源文件的虚拟路径，它应该以正斜线（/）开始，"/"表示当前 web 应用的根目录，也可以不以"/"开始。

示例如下：

```
public class PathServlet extends HttpServlet{
    publicvoid doGet(HttpServletRequest request, HttpServletResponse response)
            throwsServletException, IOException {
            ServletContext servletContext=this.getServletContext();
            String path=servletContext.getRealPath("/");
            System.out.println(path);
            String indexPath=servletContext.getRealPath("index.jsp");
            System.out.println(indexPath);
    }
}
```

2. 获取资源流

ServletContext 接口的 getResouceAsStream(String path)方法可以获取 path 指定资源的流。

参数 path 代表资源文件的虚拟路径，它应该以正斜线（/）开始，"/"表示当前 web 应用的根目录，也可以不以"/"开始。

这个方法也可以理解为，先获得资源的真实路径，再使用 InputStream input=new FileInputStream(newFile(servletContext.getRealPath(path)))创建一个输入流。

示例如下：

```
ServletContext servletContext=this.getServletContext();
//使用ServletContext获取资源流
InputStream input=servletContext.getResourceAsStream("/index.jsp");
System.out.println(input);
```

3. 获取指定目录下的所有资源路径

通过 ServletContext 接口的 getResourcePaths(String path)方法获取所有资源路径，该方法的返回值是一个 Set 集合。该方法的参数一定要以"/"开头，否则会报错。

例如，将 WEB-INF 目录下的所有资源路径都获取到：

```
ServletContext servletContext=this.getServletContext();
//使用ServletContext获取指定目录下所有资源路径
Set<String> paths=servletContext.getResourcePaths("/WEB-INF");
System.out.println(paths);//[/WEB-INF/lib, /WEB-INF/classes,/WEB-INF/web.xml]
```

注意

WEB-INF 目录下的 lib 和 classes 都是目录，但是通过 getResourcePaths()方法不会进一步获得 lib 和 classes 目录下的资源路径。

【例 6-7】 读取配置文件中数据库的信息。

jdbc.properties 文件的代码如下：

```
url = jdbc:mysql://localhost/
```

```
dbName = studb
userName = root
userPasswd = 123456
```

ContextServletTest.java 文件的代码如下：

```
public class ContextServletTest extends HttpServlet {
    protected void doPost(HttpServletRequest request, HttpServletResponse response) throws
ServletException, IOException {
        response.setContentType("text/html;charset=UTF-8");
        PrintWriter out = response.getWriter();
        ServletContext context = this.getServletContext();
        InputStream stream = context.getResourceAsStream("/WEB-INF/classes/jdbc. properties");
        Properties properties = new Properties();
        properties.load(stream);
        String dbName = properties.getProperty("dbName");
        String userName = properties.getProperty("userName");
        String userPasswd = properties.getProperty("userPasswd");
        String url = properties.getProperty("url");
        url = url + dbName + "?user=" + userName + "&password=" + userPasswd;
        try {
          Class.forName("com.mysql.jdbc.Driver");
          Connection conn = DriverManager.getConnection(url);
          Statement statement = conn.createStatement();
          String sql = "select * from student ORDER BY stuclass DESC,number ASC   ";
          ResultSet rs = statement.executeQuery(sql);
          out.println("<table border=1>");
          out.println("<th colspan=4 style=' font-size: x-large; alignment: center'> 学生信息表</th>");
          out.println(" <tr style=' text-align: center'>");
          out.println(" <td>学号</td>   <td>姓名</td>   <td>出生日期</td>   <td>所在班级</td>   </tr>");
          while (rs.next()) {
              out.println(" <tr style='text-align: center'>");
              out.println(" <td>" + rs.getString(1) + "</td>");
              out.println(" <td>" + rs.getString(2) + "</td>");
              out.println(" <td>" + rs.getString("birthday") + "</td>");
              out.println(" <td>" + rs.getString("stuclass") + "</td>");
              out.println(" </tr>");
          }
          out.println("</table></body>");
        } catch (Exception e) {
              e.printStackTrace();
          }
          out.flush();
          out.close();
    }
```

运行效果如图 6-12 所示。

图 6-12

本章小结

本章基于 Servlet 技术，首先讲解了 Servlet 的主要作用，以及如何通过 IntelliJ 工具创建并使用 Servlet 类，最后讲解了 Web 容器中初始化参数的设定和虚拟路径的配置。

1. Servlet 运行原理

web 服务器收到一个 http 请求后，判断请求内容，若是静态页面数据，则自行处理，若为动态数据，则交给 Servlet 容器，Servlet 容器找到相应 Servlet 实例处理；处理结果交给 web 服务器，再转交给客户端。

针对同一个 Servlet，Servlet 容器会在第一次收到 HTTP 请求时建立一个 Servlet 实例，然后启动一个线程，第二次收到 HTTP 请求后，Servlet 容器无须创建相同的 Servlet ，仅开启第二个线程来处理请求。

2. 编写 Servlet 的步骤

（1）实现 Servlet 接口（javax.servlet.Servlet）。

（2）重写 Servlet 接口中的 service 方法。

（3）在 WebRoot/WEB-INF/web.xml 中配置 Servlet 的路径。浏览器访问 Servlet 的路径为 web.xml。

课后练习

1. 填空题

（1）Servlet 的生命周期包括加载与实例化、__________、__________和销毁四个阶段。

（2）在编写 Servlet 时，需要继承________类，在 Servlet 中声明 doGet()和 doPost()需要____和_______类型的两个参数。

（3）在访问 Servlet 时，在浏览器地址栏中输入的路径是在______配置的。

2. 选择题

(1) Servlet 需要在 (　　) 文件中进行配置。

A. context.xml　　B. web.config　　C. web.xml　　D. webapp.xml

(2) 关于 web.xml 的配置，说法错误的是 (　　)。

A. 在 web.xml 描述中，要指定这个 Servlet 的名字

B. 在 web.xml 描述中，要指定这个 Servlet 的类

C. 在 web.xml 描述中，要为 Servlet 做 URL 映射

D. 在 web.xml 中不可以同时指定多个 Servlet

(3) 下列选项中，(　　) 可以准确地获取请求页面的一个文本框的内容

A. request.getParameter (name)　　B. request.getParameter ("name")

C. request.getParameterValues(name)　　D. request.getParameterValues("name")

(4) 假设在 myServlet 应用中有一个 MyServlet 类，在 web.xml 文件中对其进行如下配置：

```
<servlet>
    <servlet-name> mysrvlet </servlet-name>
    <servlet-class> com.wgh.MyServlet </serviet-class>
</servlet>
<servlet-mapping>
    <servlet-name> my servlet </servlet-name>
    <url-pattern> /welcome </url-pattern>
</servlet-mapping>
```

则以下选项可以访问到 MyServlet 的是 (　　)。

A. http://localhost:8080/MyServlet

B. http://localhost:8080/myservlet

C. http://localhost:8080/com/wgh/MyServlet

Dvhttp://localhost:8080/welcome

3. 简答题

(1) 简述 Servlet 是什么，以及 Servlet 的工作原理。

(2) JSP、Servlet 中的请求转发分别如何实现?

4. 编程题

(1) 使用 Servlet 类来完成一个用户登录验证的实例。即提供一个页面，让用户输入用户名和密码，这个页面提交到一个 Servlet 中，在这个 Servlet 中判断，如果用户名是“test”，密码是“123456”，就返回“用户登录成功”的信息，否则返回“用户登录失败”的信息。

(2) 在数据库中建立一个表 T_BOOK，包含图书 ID、图书名称、图书价格。编写图书模糊查询界面，输入图书名称的模糊资料，界面下方显示图书信息。要求提交给 Servlet 完成。

PART07

项目七

"天码行空"企业网站的设计与实现

企业门户网站是根据企业行业的不同来建立属于企业自己的网站，再对自己企业的产品进行宣传，树立良好的企业形象。本章将学习如何使用JSP技术访问MySQL数据库并结合HTML+CSS技术设计与实现"天码行空"企业网站。

课堂学习目标

- 系统功能分析与设计
- 后台管理系统主要功能的实现
- 企业网站主要功能的实现

素养拓展

- 坚持不懈

任务一　系统功能分析与设计

任务要求

本任务要求分析“天码行空”企业门户网站的系统功能结构和业务流程，并根据网站功能分析，进行数据库设计和主要框架设计。

任务实现

（一）系统功能结构分析

企业门户网站的建设，使企业能够展示自己的产品与特色，建立与客户更好的交流方式。企业门户网站的建设和管理水平直接影响企业形象，拥有一个美观实用的企业门户网站，已经成为大部分企业必不可少的建设内容。

企业门户网站通常由两部分组成：一部分是网站前台，用于展示企业信息，以及与客户进行交流；另一部分是网站后台，用于管理网站信息。

（1）网站前台实现首页、企业简介、公告、新闻、产品介绍和联系我们的展示页面。

（2）网站后台实现登录、新闻管理、公告管理和管理员管理等模块功能。

系统功能结构如图 7-1 所示。

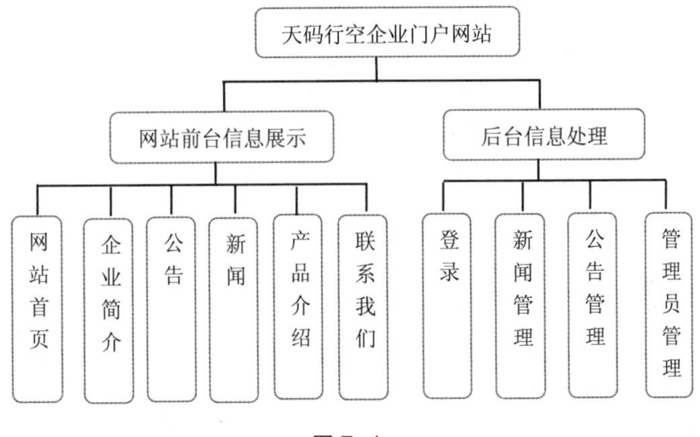

图 7-1

（二）系统业务流程

外部用户在浏览器中输入公司门户网站网址，进入网站首页，通过单击顶部的导航菜单，可以进入浏览网站首页、企业简介、公告、新闻、产品介绍、联系我们等页面。

网站管理员对门户网站显示的信息进行维护，管理员可以通过后台信息处理的登录入口，输入用户名和密码进行登录，验证成功后，进入后台管理页面，可以进行新闻管理、公告管理、用户信息管理等信息处理。

系统业务流程图如图 7-2 所示。

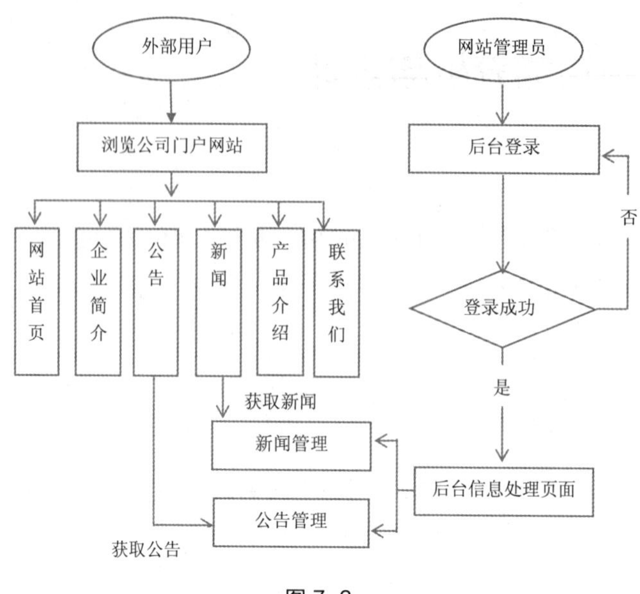

图 7-2

（三）系统开发环境

本系统的软件开发及运行环境如下。

（1）操作系统：Windows 7，Windows 8.1，Windows 10。

（2）JDK 环境：Java SE Development Kit (JDK) version 8。

（3）开发工具： IntelliJ IDEA 2018。

（4）Web 服务器：Tomcat 8.0。

（5）数据库：MySQL-5.7（配置版）。

（6）浏览器：推荐谷歌浏览器或者火狐浏览器。

（7）分辨率：推荐分辨率为 1024 像素×768 像素。

（四）系统数据库设计

根据系统功能分析，要对系统管理员用户、新闻和公告信息进行记录与维护，所以在此设计表 7-1～表 7-3 这三个表以支撑目前的系统功能，此处设计的数据库名称为 unit7website。

（1）管理员用户表(admin):主要用来保存系统管理员用户，系统管理员用户登录成功后，才可维护系统新闻和公告信息。

表 7-1　管理员用户表 admin

字段名	数据类型	是否主键	是否 Null 值	默认值	描述
AdminID	int(11)	是	否		管理员编号
AdminName	varchar (32)	否	是		用户名
AdminPwd	varchar (64)	否	是		登录密码
AdminType	Smallint(6)	否	否	0	管理员类型
LastLoginTime	Varchar(50)	否	否	'暂无登录'	登录时间

（2）企业新闻表（news）：主要用来保存企业公布的新闻，管理员在网站后台维护企业新闻后，在门户网站上可以显示最新的企业新闻。

表 7-2　企业新闻表 news

字段名	数据类型	是否主键	是否 Null 值	默认值	描述
NewsID	int(11)	是	否	自增	新闻编号
NewsTitle	varchar (60)	否	是		新闻标题
NewsContent	longtext	否	是		新闻内容
NewsTime	varchar (50)	否	否		发布时间
AdminName	Varchar(32)	否	否		管理员用户名

（3）企业公告表（notice）：主要用来保存企业公告信息，管理员在网站后台维护企业公告后，在门户网站上可以显示最新的企业公告。

表 7-3　企业公告表 notice

字段名	数据类型	是否主键	是否 Null 值	默认值	描述
NoticeId	int(11)	是	否	自增	公告编号
NoticeTitle	varchar (60)	否	是		公告标题
NoticeContent	longtext	否	是		公告内容
NoticeTime	varchar (50)	否	否		发布时间
AdminName	Varchar(32)	否	否		管理员用户名

读者可以根据自己对企业门户网站的功能设计进行企业门户网站的功能设计与扩展，并对表结构进行修改与扩展。

任务二　后台管理系统主要功能的实现

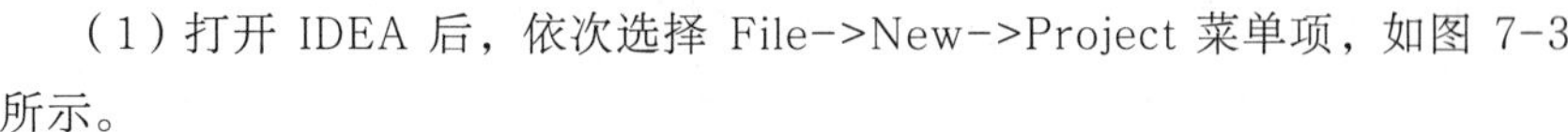

任务要求

本任务要求使用 IntelliJ IDEA（简称 IDEA）开发工具，创建门户网站后台管理项目，一步一步实现后台管理功能。

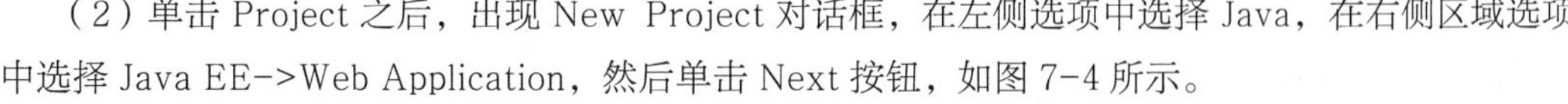

任务实现

（一）创建项目

本项目使用 IntelliJ IDEA 2018 集成开发环境（以下简称 IDEA）实现门户网站的建立，首先创建"天码行空"门户网站项目，具体步骤如下。

（1）打开 IDEA 后，依次选择 File->New->Project 菜单项，如图 7-3 所示。

（2）单击 Project 之后，出现 New Project 对话框，在左侧选项中选择 Java，在右侧区域选项中选择 Java EE->Web Application，然后单击 Next 按钮，如图 7-4 所示。

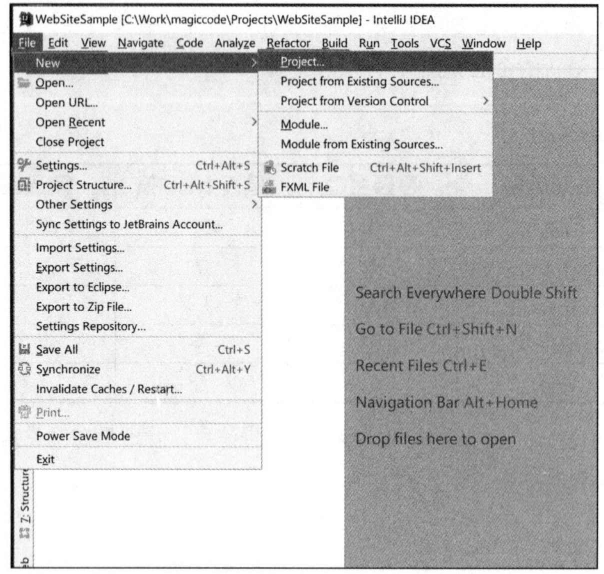

图 7-3

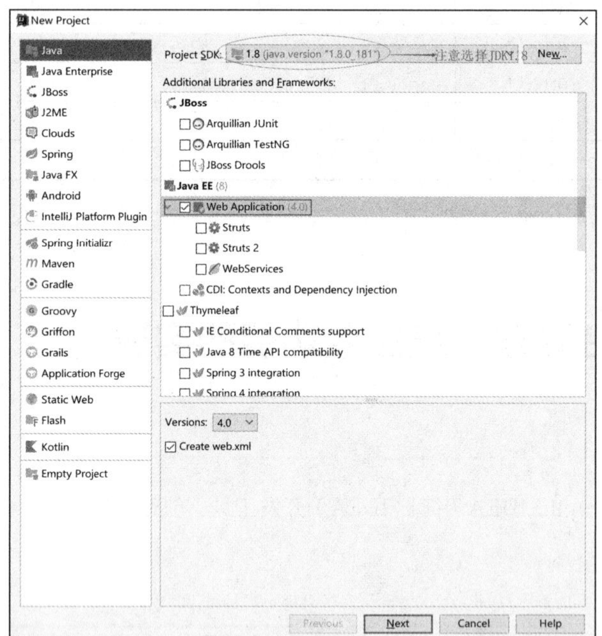

图 7-4

在 Project SDK 处，选择本机安装的 JDK 1.8 版本，如果未显示本机安装的 JDK，请查阅 IDEA 如何配置 JDK 的相关资料。

（3）选择所创建项目类型之后，请为项目命名，此处将 Project name 命名为 CodeSkyWebSite（要符合文件名命名规范），并且选择项目存储的路径，然后单击 Finish 按钮，如图 7-5 所示。

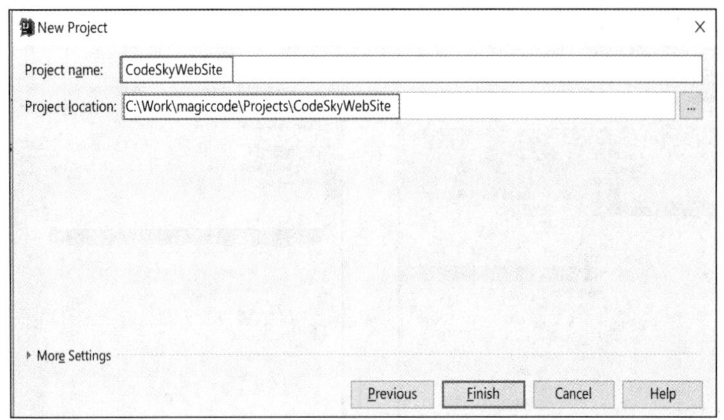

图 7-5

（4）创建 CodeSkyWebSite 项目以后，显示出项目的整体结构，一般情况下，把包和类的源代码放在 src 目录下，把网站的静态与动态网页以及网站相关资源放在 web 目录下，如图 7-6 所示。

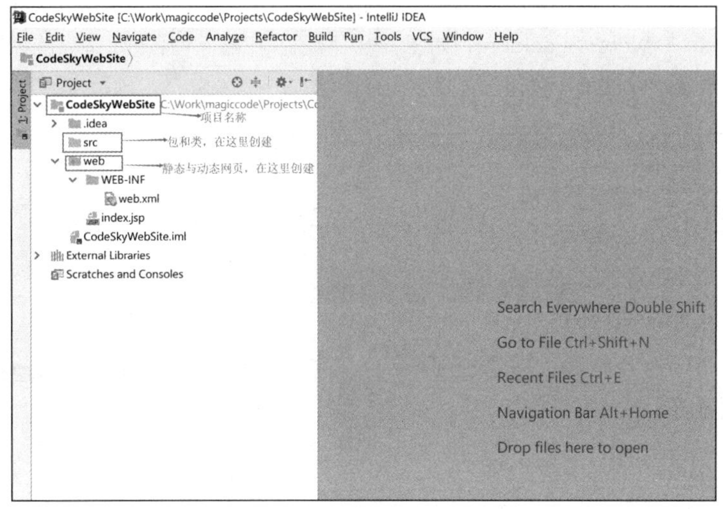

图 7-6

（二）后台登录模块的实现

后台登录模块是管理员通过后台登录页面，输入用户名与密码，连接数据库进行查询，如果用户名与密码正确就进入后台管理系统的主页面，如果输入错误就提示错误并且禁止进入后台管理系统。根据功能要求，实现后台登录模块的具体步骤如下：

创建整个网站的目录结构文件夹，右键单击 web 文件，选择 New->Directory，创建文件夹，如图 7-7 所示。分别创建后台管理系统 admin 文件夹，此文件夹内放置所有后台管理功能页面和资源；门户网站 front 文件夹，此文件夹内放置所有门户网站前台页面和资源，如图 7-8 所示。每个子系统文件夹下分别建立和引入 css（颜色、字体大小等）、img（图片）和 js（动态效果）文件夹。（css 和 js 属于静态页面开发技术，此处不再详细讲解这些代码资源，读者可以直接引入本章提供的 css、img、js 资源使用。）

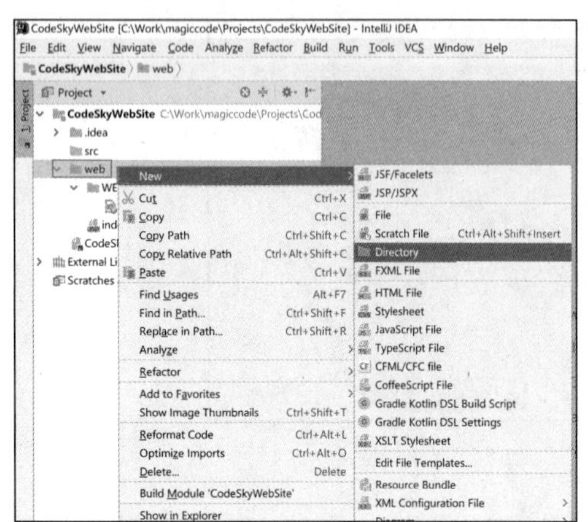

图 7-7

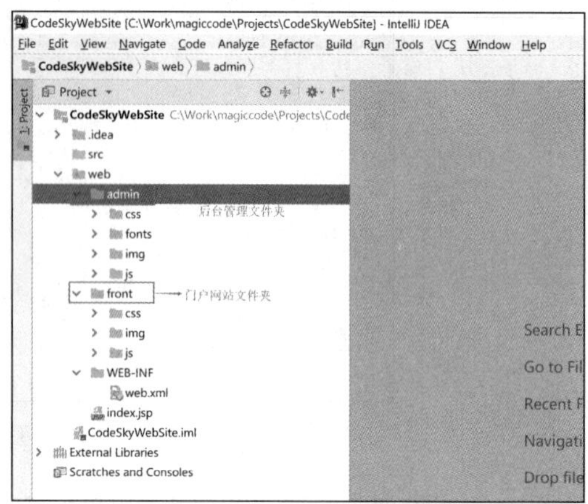

图 7-8

为了实现登录功能，要先创建数据库访问基础类、基本功能类、登录类，并且在创建类之前先创建包。右键单击 src->New->Package，如图 7-9 所示；在输入包名处，命名包名为 jspSamples.unit7.websiteSample，单击 OK 按钮，如图 7-10 所示。

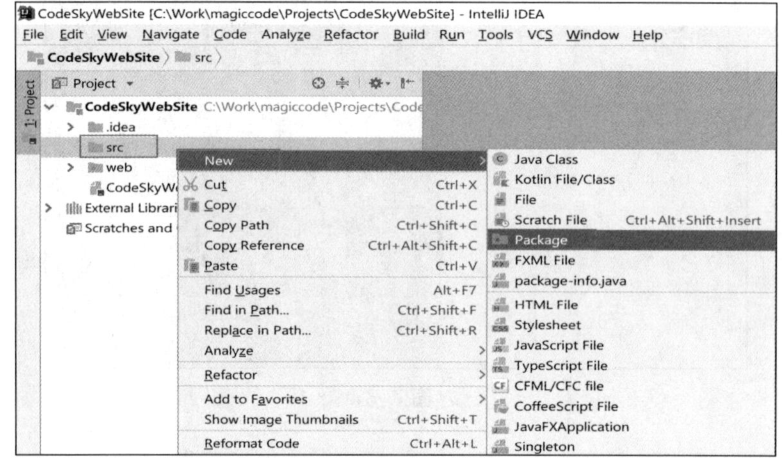

图 7-9

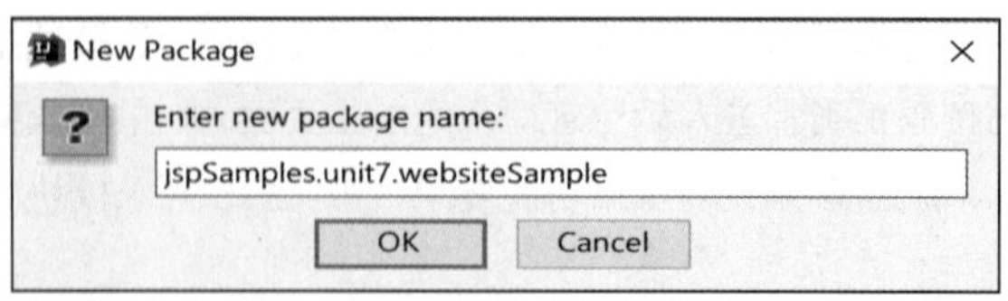

图 7-10

创建数据库访问 DBConnection 类，如图 7-11 所示，在刚创建的包处单击右键选择 New->Java Class，如图 7-12 所示，出现创建 Class 的对话框，填写类名 DBConnection，类型 Kind 选择第一个 Class。

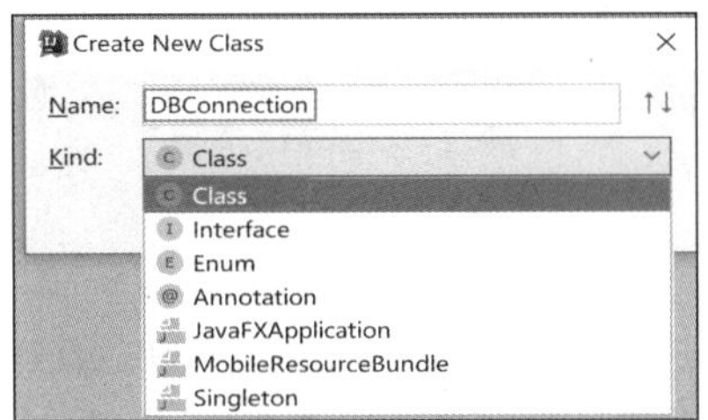

图 7-11

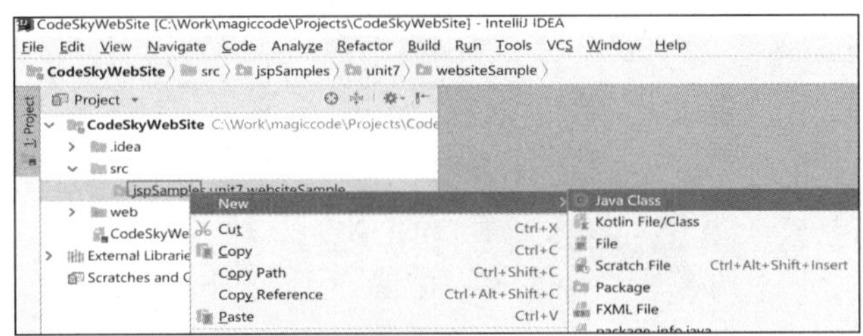
图 7-12

数据库访问 DBConnection 类主要实现数据库访问与连接，具体实现代码如下：

```
package jspSamples.unit7.websiteSample;
import java.sql.Connection;
import java.sql.DriverManager;
public class DBConnection
{

    private String FileName;
    private int DBType;
    private Connection conn;
    private String MySqlDriver;
    private String MySqlURL;
    public DBConnection()
    {
      conn = null;
    }
//根据数据库类型创建数据库连接，此处默认类型为1：mysql
 public    Connection getConn()
 {

      DBType=1;

      switch(DBType)
      {
          case 1:return(getConnToMySql());
          default:return null;
      }
 }
```

```
//创建数据库连接
        public Connection getConnToMySql()
        {
          try{
                MySqlDriver = "org.gjt.mm.mysql.Driver";
                MySqlURL = "jdbc:mysql://127.0.0.1:3306/unit7website?user=fairchild&password= huiko8213
@K&useUnicode=true&characterEncoding=UTF-8";
                Class.forName(MySqlDriver).newInstance();
                conn = DriverManager.getConnection(MySqlURL);
          }catch(Exception e){
          }
         return conn;
        }

    }
```

提示

getConn 方法中设定 DBType=1 表示此处可以设置成访问多种数据，并且用 DBType 的值来切换，值为 1 时表示为 mysql 数据库。getConnToMySql 方法中 mysql 的数据库驱动设定为：MySqlDriver = "org.gjt.mm.mysql.Driver";，数据库访问的地址设定为：MySqlURL = "jdbc:mysql://127.0.0.1:3306/unit7website?user=admin&password=123456&useUnicode=true&characterEncoding=UTF-8";，此 URL 中有 mysql 服务器的地址，数据库名称 unit7website，访问 mysql 的用户名 admin，访问 mysql 数据库的密码 123456，以及编码字符集。

数据库驱动类设定为 MySqlDriver = "org.gjt.mm.mysql.Driver"，此驱动类的压缩包为 MySQL-Driver.jar，需要将本书提供的项目资源文件中的此压缩包拷贝到“web\WEB-INF\lib”目录下。

实现“用户登录”业务逻辑，在包 jspSamples.unit7.websiteSample 下创建业务逻辑实现类 Function，在此类中实现所有的业务逻辑，所以此类会逐步实现，多次补充，此步骤主要实现用户登录业务逻辑。Function 中 CheckLogin 方法的主要实现代码如下：

```
package jspSamples.unit7.websiteSample;

import java.sql.Connection;
import java.sql.ResultSet;
import java.sql.SQLException;
import java.sql.Statement;
import java.text.SimpleDateFormat;
import java.util.Date;

public class Function {

  DBConnection DBConn = new DBConnection();
```

```
/**
     * 方法名：CheckLogin
     * 功能描述：登录验证
     * Created by 天码行空
    */
public boolean CheckLogin(Connection conn, String s1, String s2)
            throws SQLException {
        Statement stmt = conn.createStatement();
        ResultSet rs = null;
        boolean OK = true;
        String AdminPwd = "";
        String User = CheckReplace(s1);
        String Pwd = CheckReplace(s2);
        String Sql = "select * from Admin where AdminName='" + User + "'";
        rs = stmt.executeQuery(Sql);
        if (!rs.next()) {
            OK = false;
        } else {
            AdminPwd = rs.getString("AdminPwd");
            OK = Pwd.equals(AdminPwd);
        }
        return OK;

    }
}
```

提示　CheckLogin 方法的主要业务逻辑为：参数传入用户名（s1），密码（s2）和数据库连接 conn，进行数据库访问，在数据表 Admin 中查询同时匹配用户名和密码的记录，如果查询到存在记录，返回 true，否则返回 false。

创建并实现用户登录类，在包 jspSamples.unit7.websiteSample 下创建 Login 类，在此类中包含所有与登录功能相关的实现，方法的主要实现代码如下：

```
package jspSamples.unit7.websiteSample;

import java.sql.Connection;
import java.sql.SQLException;

public class Login {

  DBConnection DBConn = new DBConnection();
  Function Fun = new Function();

  public Login() {}
  public boolean LoginCheck(String s1, String s2) {
```

```
            try {
                Connection Conn = DBConn.getConn();
                boolean OK = true;
                OK = Fun.CheckLogin(Conn, s1, s2);
                return OK;

            } catch (SQLException e) {
                return false;
            }
        }
    }
```

创建用户访问页面 login.jsp，如图 7-13 所示，在 web->admin 目录下，右击选择 New->JSP/JSPX，出现创建 JSP/JSPX 页面的对话框，填写名称 login.jsp，类型 Kind 选择 JSP file，如图 7-14 所示。

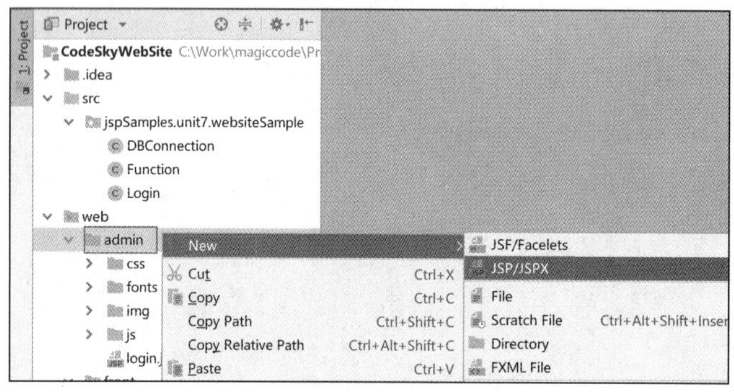

图 7-13

Create JSP/JSPX page
Name: login.jsp
Kind: JSP file
OK Cancel

图 7-14

用户登录页面 login.jsp，具体实现代码如下：

```
<%@ page language="java" import="java.util.*" pageEncoding="GBK"%>
<%@ page import="jspSamples.unit7.websiteSample.*"%>
<%
 request.setCharacterEncoding("GBK"); //设置编码方式为GBK
 String Action = request.getParameter("Action"); //判断是否单击 “登录”
 if (Action != null && Action.equals("Login")) {
     String User = request.getParameter("User"); //得到登录用户名
     String Pwd = request.getParameter("Pwd"); //得到登录密码
     out.println("<script>alert('" + User + "');</script>");
     out.println("<script>alert('" + Pwd + "');</script>");
```

```
        Login login = new Login(); //新建登录类Login
        boolean isOK = login.LoginCheck(User, Pwd); //调用方法LoginCheck，判断返回值是真还是假
        if (isOK) {
          //如果isOK=true，说明验证成功，可以进入后台页面news.jsp
          out.println("<SCRIPT  LANGUAGE='JavaScript'>alert(' 登 录 成 功 ！ ');location.href='news.jsp';
</SCRIPT>");
        } else {
          //如果isOK=false，说明验证失败，无法进入后台页面
          out.println("<SCRIPT  LANGUAGE='JavaScript'>alert(' 登 录 失 败 ！ ');location.href=  'login.jsp';
</SCRIPT>");
        }
    }
  %>
  <!DOCTYPE html>
  <html>
  <head>
  <meta charset="GBK">
  <title>后台登录-天码行空学习建站</title>
  <link rel="stylesheet" href="css/amazeui.min.css" />
  <script src="js/main.js"></script>
  </head>
  <!--引入背景图-->
  <body style="background: url(img/login-bg.png) no-repeat">
   <!--公司标题图片-->
   <div class="header" style="text-align: center; margin-top: 100px">
       <div class="am-g">
           <img src="img/loginTitle.png" />
       </div>
   </div>
   <!--登录框-->
   <div class="am-g" style="margin-top: 20px">
       <div class="am-u-lg-6  am-u-md-10  am-u-sm-centered"
           style="background: url(img/loginForm.png) no-repeat; height: 479px; width: 695px">
           <!--登录表单-->
           <form action="login.jsp" method="post" class="am-form login-form"
                 style="padding: 50px 0px 0px 120px; width: 550px"
                 onSubmit="return LoginCheck()">
                 <label for="name"> 用户名: </label> <input type="text" name="User"
                       id="User" value=""> <br> <label for="ps"> 密码:</label>
                 <input type="password" name="Pwd" id="Pwd" value=""> <br>
                 <!--登录按钮-->
                 <div class="am-cf">
                       <input name="Action" type="hidden" value="Login"> <input
                             type="submit" value="登  录" id="save"
                             style="width: 100%; border-radius: 0.5em;"
```

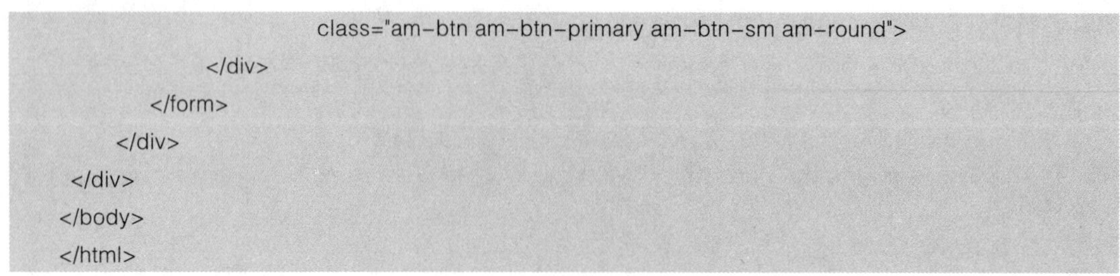

```
                    class="am-btn am-btn-primary am-btn-sm am-round">
            </div>
        </form>
    </div>
  </div>
</body>
</html>
```

配置 IDEA 下的 web 运行环境配置，首先在 IDE 环境的右上角单击下拉框，选择 Edit Configurations，如图 7-15 所示，出现项目配置对话框，如图 7-16 所示，首先单击左上角的“+”图标，在 Add New Configration 列表下找到 Tomcat Server，然后选择 Local（本地）配置。选择 Local 后，会出现配置界面，如图 7-17 所示，配置 Server 标签页，然后单击 Development 标签页，配置部署，如图 7-18 所示。配置完成后，单击 OK 按钮即可。

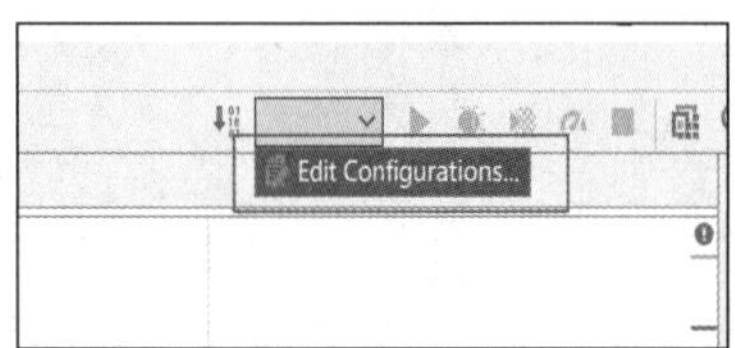

图 7-15

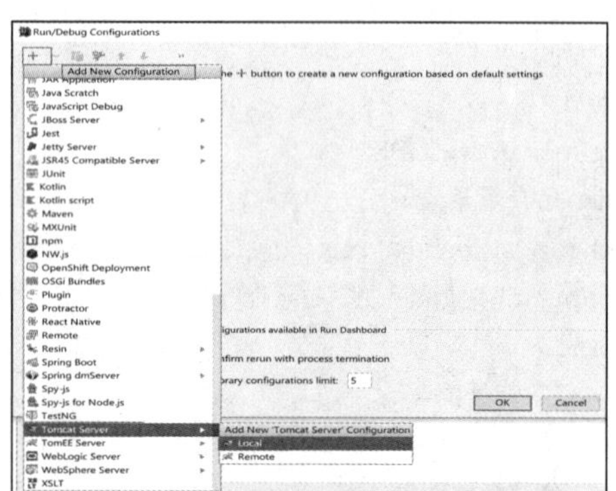

图 7-16

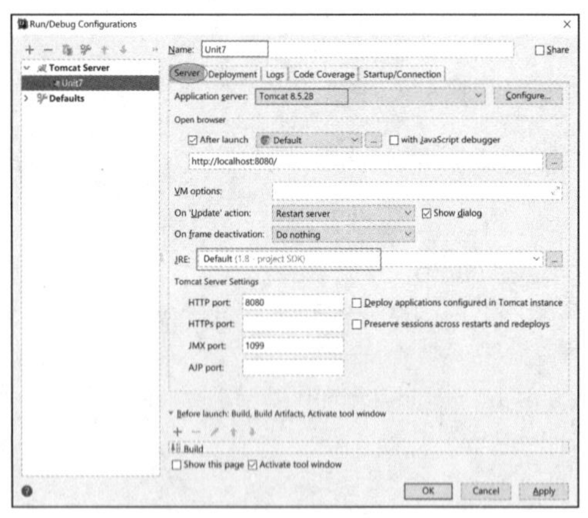

图 7-17

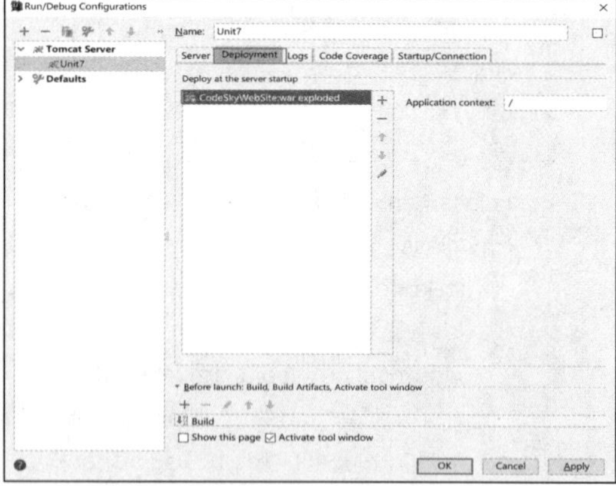

图 7-18

运行 Unit7 配置服务器，测试 login.jsp 页面。配置完成后，单击右上角 Unit7 服务器，如图 7-19 所示。

图 7-19　选择 Unit7 服务器

单击绿色运行箭头，服务器正常启动后，会根据配置的默认浏览器出现默认的 login.jsp 页面，如图 7-20 所示，网址为"http://localhost:8080/admin/login.jsp"，在用户名处输入数据库 Admin 表中已有的"admin"用户和"123456"密码，单击"登录"按钮。

图 7-20

（三）后台新闻模块的实现

后台的操作就如同产品加工厂一样，这里加工的是数据，主要包括新闻记录和公告记录的新增、删除、修改以及列表展示，通常称其为增删改查。

新闻管理模块主要是维护新闻记录，一条新闻记录由新闻标题、新闻内容、新闻作者和发布时间等要素构成，新闻记录增删改查的过程实际上也就是操作这些要素的过程。根据功能要求，后台新闻模块实现的具体步骤如下。

新闻管理界面展示，如图 7-21 所示，整个页面样式由三部分构成：顶部的 Logo 和名称，左侧的"菜单栏"和右侧的"主内容区"。

图 7-21

扩展 Function 类中的方法，实现新闻记录的增加、修改、删除、浏览、分页等功能。打开 Function 类，添加方法代码如下：

```
/**
    * 方法名：CheckReplace
```

```
    * 功能描述：字符串过滤
    * Created by 天码行空
   */
public String CheckReplace(String s) {
    try {
        if ((s == null) || (s.equals("")))
            return "";

        StringBuffer stringbuffer = new StringBuffer();
        for (int i = 0; i < s.length(); i++) {
            char c = s.charAt(i);
            switch (c) {
            case '"':
                    stringbuffer.append("&quot;");
                    break;
            case '\'':
                    stringbuffer.append("&#039;");
                    break;
            case '|':
              break;
            case '&':
                    stringbuffer.append("&amp;");
                    break;
            case '<':
                    stringbuffer.append("&lt;");
                    break;
            case '>':
                    stringbuffer.append("&gt;");
                    break;
            default:
                    stringbuffer.append(c);
            }
        }

        return stringbuffer.toString().trim();
    } catch (Exception e) {
    }
    return "";
}

/**
    * 方法名：CheckDate
    * 功能描述：日期验证
    * Created by 天码行空
   */
```

```
public String CheckDate(String[] s1, String[] s2) {
    boolean OK = true;
    StringBuffer sb = new StringBuffer();
    try {
        for (int i = 0; i < s1.length; i++) {
            if ((s1[i] == null) || (s1[i].equals(""))
                    || (s1[i].equals(" "))) {
                sb.append("<li> [ " + s2[i] + " ] 不能为空!");
                OK = false;
            }
        }
        if (OK)
            return "Yes";
        return sb.toString().trim();
    } catch (Exception e) {
    }
    return "操作失败！";
}

public String getStrCN(String s) {
    if (s == null)
        s = "";
    try {
        byte[] abyte0 = s.getBytes("GBK");
        s = new String(abyte0);
    } catch (Exception e) {
        s = "";
    }
    return s;
}

public int StrToInt(String s) {
    try {
        return Integer.parseInt(CheckReplace(s));
    } catch (Exception e) {
    }
    return 0;
}

public boolean StringToBoolean(String s) {
    return (s != null) && (s.equals("Yes"));
}

public String Page(String sPage, ResultSet rs, int intPage, int intPageSize) {
    String s = null;
```

```
int i = 0;
try {
    rs.last();

    int intRowCount = rs.getRow();
    int intPageCount;
    if (intRowCount % intPageSize == 0)
        intPageCount = intRowCount / intPageSize;
    else
        intPageCount = (int) Math.floor(intRowCount / intPageSize) + 1;
    if (intPageCount == 0)
        intPageCount = 1;

    if (intPage < 1)
        intPage = 1;
    if (intPage > intPageCount)
        intPage = intPageCount;

    if (intRowCount > intPageSize) {
        s = "<table class=\"am-table am-table-striped\" width=\"90%\"    border=\"0\" align=\"center\
" cellpadding=\"2\" cellspacing=\"0\"><tr>";
        s = s + "<td width=\"80%\" height=\"30\" class=\"chinese\"><span class=\"chinese\">";
        s = s + "当前第" + intPage + "页/共" + intPageCount+ "页,    共" +
intRowCount
                + "条记录,    " + intPageSize + "条/页";

        int showye = intPageCount;


        if (showye > 10)
            showye = 10;
        for (i = 1; i <= showye; i++)
            ;
        s = s + "</span></td>";
        s = s + "<td width=\"20%\">";
        s = s + "<table width=\"100%\" border=\"0\">";
        s = s + "<tr><td><div align=\"right\"><span class=\"chinese\">";
        s = s + "<select id=\"ipage\" name=\"ipage\" class=\"chinese\" onChange=\ "jumpMenu
('self',this,0)\">";
        s = s + "<option value=\"\" selected>请选择</option>";

        for (i = 1; i <= intPageCount; i++) {
            String sSelect = i == intPage ? "SELECTED" : "";
            s = s + "<option value=\"" + sPage + "intPage=" + i + "\""
```

```
                    + sSelect + ">第" + i + "页</option>";
            }

            s = s + "</select></span></div>";
            s = s + "</td></tr></table>";
            return s + "</td></tr></table>";
        }

        return "";
    } catch (Exception e) {
    }
    return "分页出错!";
}

public String PageFront(String sPage, ResultSet rs, int intPage,
        int intPageSize) {
    String s = null;

    int i = 0;
    try {
        rs.last();

        int intRowCount = rs.getRow();
        int intPageCount;
        if (intRowCount % intPageSize == 0)
            intPageCount = intRowCount / intPageSize;
        else
            intPageCount = (int) Math.floor(intRowCount / intPageSize) + 1;
        if (intPageCount == 0)
            intPageCount = 1;

        if (intPage < 1)
            intPage = 1;
        if (intPage > intPageCount)
            intPage = intPageCount;

        if (intRowCount > intPageSize) {
            s = "<table  width=\"90%\"  border=\"0\" align=\"left\" cellpadding=\"2\" cellspacing=\"0\">
            <tr>";
            s = s + "<td style=\"text-align:left\" width=\"80%\" height=\"30\" class=\ "chinese\"><span
            class=\"chinese\">";
            s = s + "当前第" + intPage + "页/共" + intPageCount
                    + "页,    共" + intRowCount
                    + "条记录,    " + intPageSize
                    + "条/页        ";
```

```
                int showye = intPageCount;
                if (showye > 10)
                    showye = 10;
                for (i = 1; i <= showye; i++) {
                    if (i == intPage)
                        s = s + " " + i + " ";
                    else {
                        s = s + "  <a style=\"color:#3F862E\" href=\""
                                  + sPage + "intPage=" + i + "\">" + i + "</a> ";
                    }
                }
                s = s + "</span></td>";

                return s + "</tr></table>";
            }

            return "";
        } catch (Exception e) {
        }
        return "分页出错!";
    }

    public boolean AddLog(String[] s) {
        try {
            Connection Conn = this.DBConn.getConn();
            Statement stmt = Conn.createStatement(1004, 1007);

            for (int i = 0; i < s.length; i++) {
                s[i] = getStrCN(CheckReplace(s[i]));
            }
            String sql = "insert into Log (User,LogType,LogTime,IP,Result) values (";
            sql = sql + "'" + s[0] + "',";
            sql = sql + "'" + s[1] + "',";
            sql = sql + "'" + s[2] + "',";
            sql = sql + "'" + s[3] + "',";
            sql = sql + "'" + s[4] + "')";
            stmt.executeUpdate(sql);
            stmt.close();
            Conn.close();
            return true;
        } catch (SQLException e) {
        }
        return false;
    }
```

```
public String OutError(String s) {
    try {
        StringBuffer sb = new StringBuffer();
        sb.append("<br><br><table width=\"60%\"   border=\"0\" align=\"center\" cellpadding=\"0\"
        cellspacing=\"0\">\r\n");
        sb.append("<tr><td align=\"center\" valign=\"top\">\r\n");
        sb.append("<table width=\"90%\"   border=\"1\" align=\"center\" cellpadding=\"6\" cellspacing=
        \"1\">\r\n");
        sb.append("<tr class=\"chinese\" height=\"25\"><td height=\"27\" background= \"images/ bg.
        gif\" class=\"info\">\r\n");
        sb.append("<div align=\"center\" class=\"title\">错误页面</div></td></tr>\r\n");
        sb.append("<tr class=\"chinese\" height=\"25\"><td><table cellspacing=\"4\" cellpadding=\"1\"
        >\r\n");
        sb.append("<tr><td width=\"511\" height=\"80\" align=\"middle\" valign=\"top\">\r\n");
        sb.append("<p align=\"left\"><span class=\"info1\">操作出错：</span></p><div align=\"left\"
        class=\"info1\">");
        sb.append(s + "</div></td></tr></table></td></tr>\r\n");
        sb.append("<tr><td background=\"images/bg.gif\" height=\"20\" valign=\"middle\"><div align=
        \"center\" class=\"chinese\">\r\n");
        sb.append("<a href=\"#\" onClick=\"javascript:history.go(-1)\">返回</a></div></td> </tr>
        </table></td></tr></table><br><br>\r\n");
        return sb.toString();
    } catch (Exception e) {
    }
    return "操作出错!";
}

public String OutWarn(String s) {
    try {
        StringBuffer sb = new StringBuffer();
        sb.append("<br><br><form name=\"form1\" method=\"post\" action=\"\">\r\n");
        sb.append("<table border=\"1\" align=\"center\" cellpadding=\"1\" cellspacing=\"2\">\r\n");
        sb.append("<tr><td width=\"400\" height=\"80\" align=\"middle\" valign=\"top\">\r\n");
        sb.append("<div align=\"left\" class=\"info1\">系统警告：<br><br>\r\n");
        sb.append("    ");
        sb.append(s);
        sb.append("</div></td></tr>\r\n");
        sb.append("<tr><td height=\"20\" align=\"middle\" valign=\"top\"><div align=\"center\">\r\n");
        sb.append("<input name=\"Submit\" type=\"button\" class=\"button\" value=\"取消\"
        onClick=\"javascript:history.go(-1);\">  \r\n");
        sb.append("<input name=\"OK\" type=\"hidden\" id=\"OK\" value=\"Yes\">\r\n");
        sb.append("<input name=\"Submit2\" type=\"submit\" class=\"button\" value=\"确定\">\r\n");
        sb.append("</div></td>\r\n");
        sb.append("</tr></table></form>\r\n");
        return sb.toString();
```

```
        } catch (Exception e) {
        }
        return "操作出错!";
    }

    //获取新闻列表
    public StringBuffer ListNews(StringBuffer sb, ResultSet rs, String strPage,
            String sPage) throws SQLException {

        int i;
        int intPage = 1;
        int intPageSize = 5;

        if (!rs.next()) {
            sb.append("<tr height=\"25\" bgcolor=\"#d6dff7\"   class=\"info1\"><td colspan=\"4\">\r\n");
            sb.append("<div align=\"center\"><b>没有记录！</b></div></td></tr>\r\n");
        } else {

            intPage = StrToInt(sPage);
            sPage = CheckReplace(strPage);
            if (intPage == 0)
                intPage = 1;

            rs.absolute((intPage - 1) * intPageSize + 1);
            i = 0;
            while (i < intPageSize && !rs.isAfterLast()) {
                int NewsID = rs.getInt("NewsID");
                String NewsTitle = rs.getString("NewsTitle");
                String NewsContent = rs.getString("NewsContent");
                String NewsTime = rs.getString("NewsTime");
                String AdminName = rs.getString("AdminName");

                sb.append("<tr>");
                sb.append("<td class=\"table-id\">" + NewsID + "</td>");

                sb.append("<td>" + NewsTitle + "</td>");
                sb.append("<td class=\"table-title\">" + AdminName + "</td>");
                sb.append("<td class=\"table-title\">" + NewsTime + "</td>");
                sb.append("<td><div class=\"am-btn-toolbar\">");
                sb.append("<div class=\"am-btn-group am-btn-group-xs\">");
                sb.append("<input type=\"hidden\" value=\"" + NewsID + "\">");
                sb.append("<input type=\"hidden\" value=\"" + NewsContent
                        + "\">");
                sb.append("<input type=\"hidden\" value=\"" + NewsTitle + "\">");
                sb.append("<a style=\"background:#2167A9\" onclick=\"edit(this);\"");
```

```
                sb.append("class=\"am-btn am-btn-primary am-btn-xs \"");
                sb.append("href=\"javascript:void(0);\"> <span></span>修改<a> ");
                sb.append("<a rel=\""
                        + NewsID
                        + "\" onclick=\"del(this);\" class=\"am-btn am-btn-warning am-btn-xs \"
                        href=\"javascript:void(0);\"> "
                        + "<span></span>删除<a>");
                sb.append("</div></div></td></tr>");
                rs.next();
                i++;
            }
            sb.append(Page(sPage, rs, intPage, intPageSize));
        }
        return sb;
    }

//添加新闻信息
    public String AddNews(Connection Conn, Statement stmt, ResultSet rs,
            String[] s) throws SQLException {

        int z = 0;
        int newNum = 0;

        if (!rs.next()) {
            newNum = 1;
        } else {
            while (z < 1 && !rs.isAfterLast()) {
                int NewsID = rs.getInt("NewsID");
                newNum = NewsID + 1;
                break;
            }
        }

        for (int i = 0; i < s.length; i++) {
            if (i != 1)
                s[i] = getStrCN(CheckReplace(s[i]));
            else
                s[i] = getStrCN(s[i]);
        }

        SimpleDateFormat format1 = new SimpleDateFormat("yyyy-MM-dd HH:mm");
        String newsTime = format1.format(new Date());

        StringBuffer sql = new StringBuffer();
        sql.append("insert into News (NewsID,NewsTitle,NewsContent,NewsTime,AdminName) values ("
```

```
                + " '"
                + newNum
                + "',"
                + " '"
                + s[0]
                + "',"
                + " '"
                + s[1]
                + "'," + " '" + newsTime + "'," + " 'mr')");
        try {

            Conn.setAutoCommit(false);
            stmt.execute(sql.toString());
            Conn.commit();
            Conn.setAutoCommit(true);
            stmt.close();
            Conn.close();

            return "Yes";
        } catch (Exception e) {
            Conn.rollback();
            e.printStackTrace();
            Conn.close();
            return "添加成功!";
        }
    }

    //删除新闻信息
    public boolean DelNews(Connection Conn, Statement stmt, int NewsID)
                throws SQLException {
        if (NewsID == 0)
                return false;
        else {
                try {
                    String sql = "delete from News where NewsID=" + NewsID;

                    Conn.setAutoCommit(false);
                    stmt.executeUpdate(sql);

                    Conn.commit();
                    Conn.setAutoCommit(true);

                    stmt.close();
                    Conn.close();
                    return true;
```

```
            } catch (Exception e) {
                Conn.rollback();
                // e.printStackTrace();
                Conn.close();
                return false;
            }
        }
    }

//修改新闻信息
  public String EditNews(Connection Conn, Statement stmt, String[] s,
            String newsId) throws SQLException {
        for (int i = 0; i < s.length; i++) {
            s[i] = getStrCN(CheckReplace(s[i]));
        }

        int NewsID = StrToInt(newsId);

        StringBuffer sql = new StringBuffer();
        sql.append("update News set NewsTitle='" + s[0] + "'"
                + " ,NewsContent='" + s[1] + "'" + " where NewsID='" + NewsID
                + "'");

        stmt.executeUpdate(sql.toString());
        stmt.close();
        Conn.close();

        return "Yes";
    }

//获取新闻信息列表（前台）
  public StringBuffer ListNewsFront(StringBuffer sb, ResultSet rs,
            String toPage, String pageNum) throws SQLException {
        int i;
        int intPage = 1;
        int intPageSize = 5;
        if (!rs.next()) {
            sb.append("<tr height=\"25\" bgcolor=\"#d6dff7\"   class=\"info1\"><td colspan=\"5\">\r\n");
            sb.append("<div align=\"center\"><b>没有记录!</b></div></td></tr>\r\n");
        } else {

            intPage = StrToInt(pageNum);
            toPage = CheckReplace(toPage);
            if (intPage == 0)
                intPage = 1;
```

```
            rs.absolute((intPage - 1) * intPageSize + 1);
            i = 0;
            while (i < intPageSize && !rs.isAfterLast()) {
                int NewsID = rs.getInt("NewsID");
                String NewsTitle = rs.getString("NewsTitle");
                String NewsTime = rs.getString("NewsTime");
                String AdminName = rs.getString("AdminName");

                sb.append("<tr>");
                sb.append("<td>" + NewsTitle + "</td>");
                sb.append("<td style=\"text-align:center\">" + AdminName + "</td>");
                sb.append("<td style=\"text-align:center\">" + NewsTime + "</td>");
                sb.append("<td style=\"text-align:center\"><a style=\"color:#3F862E\" target=\"_blank\"
                href=\"newsFrontDetail.jsp?newsId="
                         + NewsID + "\">详情</a></td></tr>");

                rs.next();
                i++;
            }
            sb.append(PageFront(toPage, rs, intPage, intPageSize));
        }
        return sb;
    }

    //获取每条新闻详细信息（前台）
    public StringBuffer FrontNewsDetail(StringBuffer sb, ResultSet rs)
            throws SQLException {
        int i = 0;
        while (i < 1 && !rs.isAfterLast()) {
            rs.next();
            String NewsTitle = rs.getString("NewsTitle");
            String NewsContent = rs.getString("NewsContent");

            String[] content = NewsContent.split("#");

            sb.append("<br><h2 style=\"font-size:28px;margin-left:30px\">"
                     + NewsTitle + "</h2>");

            for (int j = 0; j < content.length; j++) {
                sb.append("<p>" + content[j] + "</p>");
            }
            rs.next();
            i++;
        }
```

```
        return sb;
    }
```

创建并实现新闻管理类，在包 jspSamples.unit7.websiteSample 下创建 News 类，在此类中包含所有与新闻管理功能相关的实现，方法的主要实现代码为：

```
package jspSamples.unit7.websiteSample;

import java.sql.Connection;
import java.sql.ResultSet;
import java.sql.Statement;

/**
 * 文件名：News.java
 * 文件功能描述：新闻管理模块的增删改查功能操作
*/
public class News {

  DBConnection DBConn = new DBConnection();           //引入数据库连接
  Function Fun = new Function();                      //引入功能命令，如验证密码、页面渲染等

  /**
       * 方法名：ListNews
       * 功能描述：实现新闻列表页（查）
       * @param toPage:分页跳转页面地址，pageNum:每页显示记录数量
       * @return 数据库查询，新闻列表字符串结果集
       * Created by 天码行空
      */
  public String ListNews(String toPage, String pageNum) {
        try {
            Connection Conn = DBConn.getConn();                       //获取数据库
            Statement stmt = Conn.createStatement();                  //创建数据库连接状态
            ResultSet rs = null;                                      //创建结果集查询
            StringBuffer resultData = new StringBuffer();             //创建结果字符串集合
            String sSql = "select * from News order by NewsID desc";  //创建数据库查询语句
            rs = stmt.executeQuery(sSql);                             //执行数据库查询
            resultData=Fun.ListNews(resultData,rs,toPage,pageNum);    //获取返回分页结果
            rs.close();                                               //关闭结果集查询连接
            stmt.close();                                             //关闭数据库连接
            Conn.close();
           return resultData.toString();                              //给前台返回结果集
        } catch (Exception e) {
            return "No";                                       //若获取数据库失败，返回“No”
        }
  }

  /**
```

```
    * 方法名：AddNews
    * 功能描述：新增新闻数据记录（增）
    * @param newsData:需要提交的新闻数据
    * @return 返回新增后的状态。返回“yes”，说明新增成功，否则新增失败
    * Created by 天码行空
    */
public String AddNews(String [] newsData)
    {
    try
     {
          Connection Conn = DBConn.getConn();                      //获取数据库
     Statement stmt = Conn.createStatement();                      //创建数据库连接状态
     ResultSet rs = null;                                          //创建结果集查询
     String sSql = "select * from News order by NewsID desc";      //创建数据库查询语句
     rs = stmt.executeQuery(sSql);                                 //执行数据库查询
     String result=Fun.AddNews(Conn,stmt,rs,newsData);             //获取查询结果
     return result;                                                //返回result 结果
    }catch(Exception e){
          return "添加失败";                                  //若获取数据库失败，返回“添加失败”
      }
    }

/**
    * 方法名：DelNews
    * 功能描述：删除单条新闻数据（删）
    * @param newsId:新闻唯一主键
    * @return 返回删除后的状态。返回true，说明删除成功，否则删除失败
    * Created by 天码行空
    */
public boolean DelNews(String newsId)
    {
     try{
     Connection Conn = DBConn.getConn();                 //获取数据库
     Statement stmt = Conn.createStatement();            //创建数据库连接状态
     int NewsID = Fun.StrToInt(newsId);                  //将字符串newsId转换为数字类型NewsID
     return Fun.DelNews(Conn,stmt,NewsID);               //删除数据并返回删除结果
     }catch(Exception e){
          return false;
     }
}

/**
    * 方法名：EditNews
    * 功能描述：修改单条新闻数据（改）
    * @param newsId:新闻唯一主键;newsData:修改数据
```

```
    * @return 返回修改后的状态。返回"yes"，说明修改成功，否则修改失败
    * Created by 天码行空
   */
public String EditNews(String[] newsData, String newsId) {
    try {
        Connection Conn = DBConn.getConn();                    //获取数据库
        Statement stmt = Conn.createStatement();               //创建数据库连接状态
        return Fun.EditNews(Conn,stmt,newsData,newsId);  //修改操作，并且返回状态
    } catch (Exception e) {
        return "数据库连接失败!";
    }
}


/**
    * 方法名：ListNewsFront
    * 功能描述：前台新闻列表页面
    * @param toPage:分页跳转页面地址，pageNum:每页显示记录数量
    * @return 数据库查询，新闻列表字符串结果集
    * Created by 天码行空
   */
public String ListNewsFront(String toPage,String pageNum)
   {
    try
    {
        Connection Conn = DBConn.getConn();
    Statement stmt = Conn.createStatement();
    ResultSet rs = null;
            StringBuffer resultData = new StringBuffer();
            String sSql = "select * from News order by NewsID desc";
    rs = stmt.executeQuery(sSql);
    resultData=Fun.ListNewsFront(resultData,rs,toPage,pageNum); //获取返回分页结果
        rs.close();
        stmt.close();
        Conn.close();
        return resultData.toString();
    }catch(Exception e)
     {
        return "No";
     }
   }
/**
    * 方法名：FrontNewsDetail
    * 功能描述：前台新闻详情页面
    * @param id:文章记录唯一主键
```

```
        * @return 数据库查询，新闻详情字符串结果集
        * Created by 天码行空
        */
    public String FrontNewsDetail(String id) {
        try {
            Connection Conn = DBConn.getConn();
            Statement stmt = Conn.createStatement();
            ResultSet rs = null;
            int NewsID = Fun.StrToInt(id);
            if (NewsID == 0)
                return "No";
            else {
                try {
                    String sql = "select * from News where NewsID=" + NewsID;
                    rs = stmt.executeQuery(sql);
                    StringBuffer sb = new StringBuffer();
                    sb=Fun.FrontNewsDetail(sb, rs);
                    rs.close();
                    stmt.close();
                    Conn.close();
                    return sb.toString();
                } catch (Exception e) {
                    Conn.rollback();
                    Conn.close();
                    return "No";
                }
            }
        } catch (Exception e) {
            return "No";
        }

    }
}
```

实现新闻列表显示页面，在 web->admin 目录下新建 new.jsp，此页面用于显示新闻列表，实现代码如下：

```
<%@ page language="java" import="java.util.*" pageEncoding="GBK"%>
<%@ page import="jspSamples.unit7.websiteSample.*"%>
<%
 request.setCharacterEncoding("GBK"); //设置编码方式为GBK
%>
<!doctype html>
<html>
<head>
<meta charset="GBK">
<title>天码行空学习建站</title>
```

```
<link rel="stylesheet" href="css/amazeui.min.css"/>
<link rel="stylesheet" href="css/admin.css"/>
<script src="js/main.js"></script>
<script src="js/news.js"></script>
</head>
<body style="background-color:#CDDBE8;">
 <header class="am-topbar admin-header" style="background-color:#2167A9;height:60px">
      <div class="am-topbar-brand" style="color:white;font-size:20px">
           <img src="img/logo1.png" width="50px"/><strong>  天码行空学习建站</strong>
           <small>后台管理</small>
      </div>
      <div class="am-collapse am-topbar-collapse"   id="topbar-collapse">
           <ul
                class="am-nav am-nav-pills am-topbar-nav am-topbar-right admin-header-list">
                <li class="am-dropdown" ><a class="am-dropdown-toggle" style="color:white"
                     href="login.jsp"> 退出  </a></li>
           </ul>
      </div>
 </header>

 <div class="am-cf admin-main" style="height:760px;background:#CDDBE8;">
     <!-- 左侧菜单 -->
     <div class="admin-sidebar am-offcanvas" style="background-color:#859FCD;margin:10px 10px"
     id="admin-offcanvas">
           <div class="am-offcanvas-bar admin-offcanvas-bar" style="height:200px">
                <ul class="am-list admin-sidebar-list" >
                     <li style="background-color:#859FCD;"><a href="news.jsp" style="color:white"
                     title="新闻管理">
                     <img src="img/title1.png" width="25px"/> 新闻管理</a></li>
                     <li style="background-color:#859FCD;"><a href="notice.jsp" style="color:white"
                     title="公告管理">
                     <img src="img/title2.png" width="25px"/> 公告管理</a></li>
                     <li style="background-color:#859FCD;"><a href="adminUser.jsp" style="color:
                     white"title="管理员管理">
                     <img src="img/title3.png" width="25px"/> 管理员管理</a></li>
                </ul>

           </div>
      </div>

      <!-- 新闻列表 -->
      <div class="admin-content" style="margin:10px 10px">
           <div class="am-cf am-padding">
                <div class="am-fl am-cf">
                     <strong>后台管理</strong>/ <strong>新闻管理</strong>
                </div>
```

```
            </div>
            <div class="am-g">
                <div class="am-u-sm-12 am-u-md-6">
                    <div class="am-btn-toolbar">
                        <div class="am-btn-group am-btn-group-xs">
                            <button type="button" class="am-btn am-btn-warning"
                                data-am-modal="{target: '#new-popup'}">
                                <span class="am-icon-plus"></span> 添加新闻
                            </button>
                        </div>
                    </div>
                </div>
            </div>

            <div class="am-g" style="height: 300px">
                <div class="am-u-sm-12">
                    <form class="am-form">
                        <table class="am-table am-table-striped am-table-hover table-main">
                            <thead>
                                <tr>
                                    <th class="table-id">序号</th>
                                    <th class="table-title">新闻标题</th>
                                    <th class="table-title">创建人</th>
                                    <th class="table-author">创建时间</th>
                                    <th class="table-author">操作</th>
                                </tr>
                            </thead>
                            <tbody>
<%
 request.setCharacterEncoding("GBK");
 News news = new News();
 String pageNum = request.getParameter("intPage");                //获取每页显示记录的数量
 String toPage = request.getContextPath() + request.getServletPath()+ "?";
//获取跳转页面的地址
 String sOK = news.ListNews(toPage, pageNum);//调用方法ListNews，获取后台返回的页面结果
 if (sOK.equals("No")) {
        out.println("数据服务器出现错误!");  //返回“No”，说明返回失败
 } else {
        out.println(sOK);                //表示成功，渲染出结果值
 }
%>
                            </tbody>
                        </table>
                    </form>
                </div>
```

```
            </div>
        </div>
</div>
<footer>
        <hr>
        <p style="text-align: center;color:#2167A9" class="am-padding-left">天码行空学习建站</p>
</footer>

<!-- 新增新闻 -->
<div class="am-popup" id="new-popup">
        <div class="am-popup-inner">
            <div class="am-popup-hd">
                <h4 class="am-popup-title">
                    添加新闻
                </h4>
                <span data-am-modal-close class="am-close">&times;</span>
            </div>

            <div class="am-popup-bd">
                <!-- 提交新闻数据记录 -->
                <form action="newsAdd.jsp" method="post"
                    class="am-form" id="new-msg">
                    <fieldset>
                        <div class="am-form-group">
                            <label for="doc-vld-ta-2-1">
                                新闻标题：
                            </label>
                            <input name="NewsTitle" type="text" maxlength="32"
                                placeholder="请输入新闻标题" data-validation-message= "不能为
                                空" required />
                        </div>
                        <div class="am-form-group">
                            <label for="doc-vld-ta-2-1">
                                新闻内容：
                            </label>
                            <textarea name="NewsContent" cols="30" rows="10"
                                placeholder="请输入新闻内容。段落间请用#分隔。" data-validation-
                                message="不能为空" required></textarea>
                        </div>

                        <input name="Action" type="hidden" value="Add">

                        <button class="am-btn am-btn-secondary" type="submit">
                            提交
                        </button>
```

```
                                <button onclick='$("#new-popup").modal("close");'
                                    class="am-btn am-btn-secondary" type="button">
                                    关闭
                                </button>
                            </fieldset>
                        </form>
                        <!-- 提交新闻数据记录结束 -->
                    </div>

                </div>
            </div>
            <!-- 新增新闻结束 -->
            <!-- 删除新闻开始 -->
            <div class="am-modal am-modal-confirm" tabindex="-1" id="my-confirm">
                <div class="am-modal-dialog">
                    <div class="am-modal-bd">
                        确定要删除当前新闻吗?
                    </div>
                    <div class="am-modal-footer">
                        <span class="am-modal-btn" data-am-modal-cancel>取消</span>
                        <span class="am-modal-btn" data-am-modal-confirm>确定</span>
                    </div>
                </div>
            </div>
            <!-- 删除新闻结束 -->
            <!-- 修改新闻开始 -->
            <div class="am-popup" id="edit-popup">
                <div class="am-popup-inner">
                    <div class="am-popup-hd">
                        <h4 class="am-popup-title">
                            修改新闻
                        </h4>
                        <span data-am-modal-close class="am-close">&times;</span>
                    </div>

                    <div class="am-popup-bd">
                        <form action="newsEdit.jsp" method="post"
                            class="am-form" id="edit-msg">
                            <fieldset>
                                <div class="am-form-group">
                                    <label for="doc-vld-ta-2-1">
                                        新闻标题：
                                    </label>
                                    <input id="upd_NewsTitle" name="upd_NewsTitle" type="text" maxlength
                                    ="32"
```

```
                placeholder="请输入新闻标题" data-validation-message= "不能为
                空" required />
            </div>
            <div class="am-form-group">
                <label for="doc-vld-ta-2-1">
                    新闻内容：
                </label>
                <textarea id="upd_NewsContent" name="upd_NewsContent"  cols="30"
                rows="10"
                    placeholder="请输入新闻内容" data-validation-message="不能为
                    空" required></textarea>
            </div>

            <input name="Action" type="hidden" value="Edit">
            <input id="newsId" name="newsId" type="hidden" value="">

            <button class="am-btn am-btn-secondary" type="submit">
                提交
            </button>
            <button onclick='$("#edit-popup").modal("close");'
                class="am-btn am-btn-secondary" type="button">
                关闭
            </button>
          </fieldset>
        </form>
      </div>
    </div>
  </div>
  <!-- 修改新闻结束 -->
</body>
</html>
```

新闻列表页面如图 7-22 所示，主要包括“添加新闻”按钮、“修改”按钮、“删除”按钮，新闻列表，翻页下拉框。

后台管理/ 新闻管理

+ 添加新闻

序号	新闻标题	创建人	创建时间	操作
8	测试新闻2	mr	2018-08-16 22:58	修改 删除
7	新的测试新闻1	mr	2018-08-13 01:40	修改 删除
6	JavaWeb编程词典个人版新品上架!	mr	2016-10-10 10:05	修改 删除
5	JSP编程词典第一次升级	mr	2016-10-10 10:01	修改 删除
4	祝全国读者端午节快乐!	mr	2016-10-10 10:00	修改 删除

当前第1页/共2页，共8条记录，5条/页　　第1页

图 7-22

实现添加新闻表单页面，在 web->admin 目录下新建 newsAdd.jsp，此页面用于获取填写的新

闻信息条目，并且调用 News 新闻业务类，把新闻信息添加到数据库中，实现代码如下（此处为了降低实现难度使用 JSP 页面取得页面数据，读者也可以使用 servlet 技术取得页面数据）：

```
<%@ page contentType="text/html; charset=GBK" language="java" %>
<%@ page import="jspSamples.unit7.websiteSample.*"%>
<%
 request.setCharacterEncoding("GBK");
 News news = new News();
 String Action = request.getParameter("Action");

 if (Action!=null && Action.equals("Add"))
 {
     String [] s = new String[2];                          //创建字符串数组
     s[0] = request.getParameter("NewsTitle");             //获取新闻标题
     s[1] = request.getParameter("NewsContent");           //获取新闻内容
     String result = news.AddNews(s);                      //将新闻记录数据提交给后台
     if (result.equals("Yes"))                             //根据返回的结果判断页面走向
     {
         out.print("<script>alert('添加新闻成功!');location.href='news.jsp';</script>");
         return;
     }
       else
     {
         out.print("<script>alert('添加新闻失败!');location.href='news.jsp';</script>");
         return;
     }
 }
```

添加新闻页面表单如图 7-23 所示，主要包括“新闻标题”输入框、“新闻内容”文本框、“提交”按钮和“关闭”按钮。

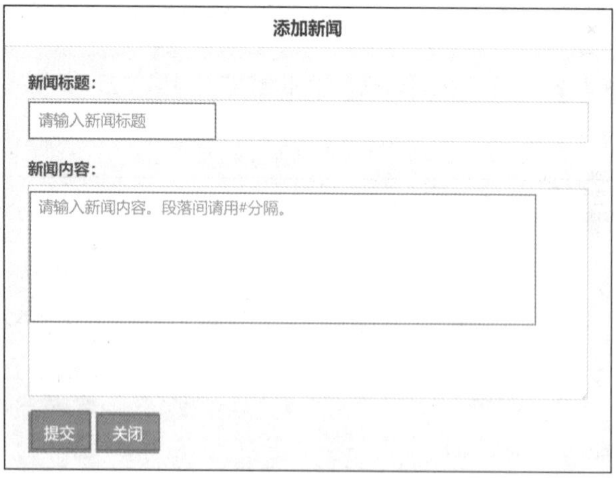

图 7-23

实现修改新闻表单页面，在 web->admin 目录下新建 newsEdit.jsp，此页面用于获取填写的新闻信息条目，并且调用 News 新闻业务类，在数据库中找到需要修改的记录并且进行内容修改，实现代码如下（此处为了降低实现难度使用了 JSP 页面取得页面数据，读者也可以使用 servlet 技术取得页面数据）：

```
<%@ page contentType="text/html; charset=GBK" language="java" %>
<%@ page import="jspSamples.unit7.websiteSample.*"%>
<%
 request.setCharacterEncoding("GBK");                    //设置编码方式为GBK
 News News1 = new News();
 String NewsID = request.getParameter("newsId");
 String Action = request.getParameter("Action");
 if (Action!=null && Action.equals("Edit"))
 {
     String [] s = new String[2];
     s[0] = request.getParameter("upd_NewsTitle");
     s[1] = request.getParameter("upd_NewsContent");

     String sOK = News1.EditNews(s,NewsID);
     if (sOK.equals("Yes")){
      out.println("<script>alert('修改新闻成功!');location.href='news.jsp';</script>");
        return;
     }
     else {
      out.println("<script>alert('修改新闻失败!');location.href='news.jsp';</script>");
        return;
     }
 }
%>
```

实现删除新闻信息页面，在 web->admin 目录下新建 newsDel.jsp，此页面用于获取填写的新闻信息条目，并且调用 News 新闻业务类，在数据库中找到需要删除的记录并且进行确认删除，实现代码如下（此处为了降低实现难度使用 JSP 页面取得页面数据，读者也可以使用 servlet 技术取得页面数据）：

```
<%@ page contentType="text/html; charset=GBK" language="java"%>
<%@ page import="jspSamples.unit7.websiteSample.*"%>
<%
 request.setCharacterEncoding("GBK"); //设置编码方式为GBK
%>
<%
 News news = new News();
 String NewsID = request.getParameter("NewsID");              //获取新闻记录唯一主键
 if (news.DelNews(NewsID))                                    //将数据提交给后台，获取返回值
     out.print("<script>alert('删除新闻成功!');location.href='news.jsp';</script>");
 else {
     out.print("<script>alert('删除新闻失败!');location.href='news.jsp';</script>");
 }
%>
```

提示　按照如上步骤已经实现"新闻管理"模块，"公告管理"和"用户管理"功能实现与新闻管理模块实现的基本方式相同，此处不再赘述，留给读者自行练习和实现，详细代码可参见本章示例代码（7.2.3_新闻管理模块实现）。

任务三　企业网站主要功能的实现

任务要求

本任务要求实现"天码行空"门户网站的整体展示，主要技术点为把后台维护的新闻信息、公告信息等最终展示在企业门户网站上，分为网站静态页面内容(首页、企业简介、产品介绍、联系我们)和网站动态数据页面（新闻、公告）。

任务实现

（一）创建前台网站项目结构

首先在原项目结构的基础上，创建前台门户网站目录结构，如图 7-24 所示，在 web 目录下创建 front 文件夹，作为门户网站前台文件夹。

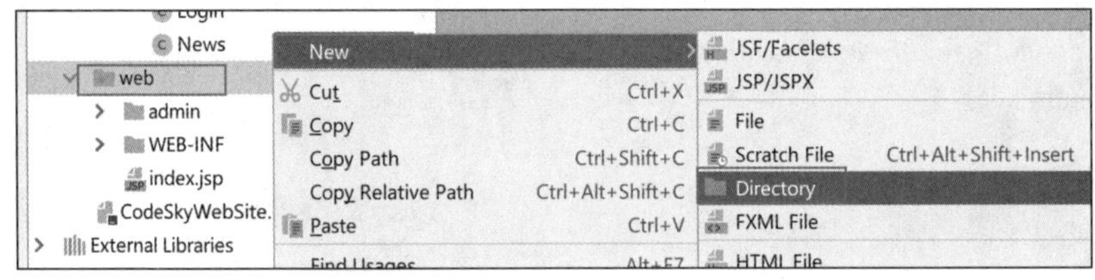

图 7-24

引入网站资源文件：front 文件夹下分别建立和引入 css（颜色、字体大小等）、img（图片）和 js（动态效果）文件夹。（css 和 js 属于静态页面开发技术，此处不再详细讲解这些代码资源，读者可以直接引入本项目提供的 css、img、js 资源），如图 7-25 所示。

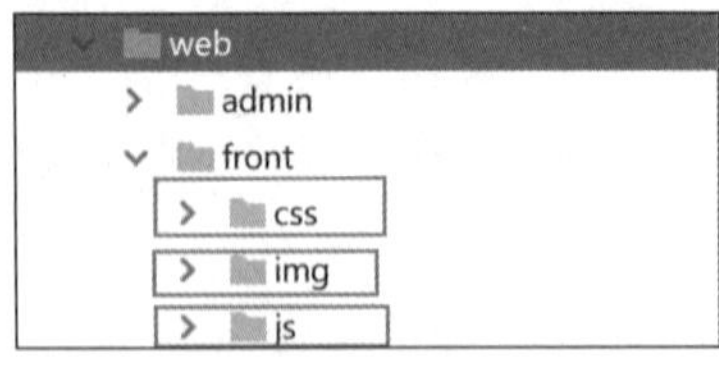

图 7-25

（二）网站主要静态页面的实现

静态网页仅仅使用 HTML、CSS 和 JavaScript 就可以制作出来，简单易学。本门户网站在首页、企业简介、产品介绍、联系我们等处使用了静态页面技术，以下着重介绍首页的实现过程。

首页是网站的默认入口，一般都是从 index.html 文件开始创建编码，所以 index.html 约定俗成地命名为网站首页文件，先在 front 文件夹下创建 index.html，如图 7-26 所示。

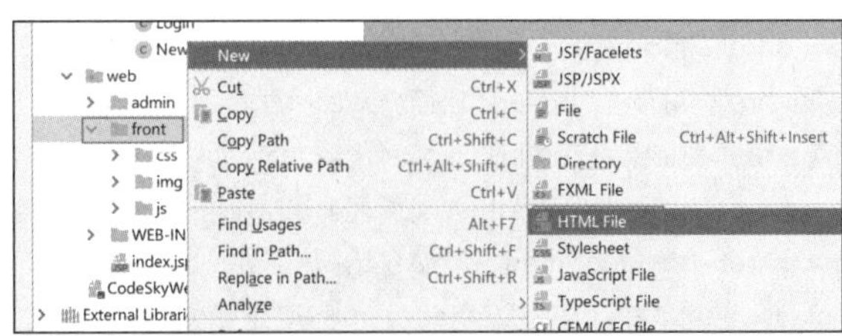

图 7-26

首页布局为三大部分：顶层的 Logo 和菜单；中间区域的企业形象图；底部区域的主要功能导航和地址电话，如图 7-27 所示。

图 7-27

网站首页实现的主要代码如下：

```
<!DOCTYPE html>
<html>
<!--网站头部-->
<head>
<meta charset="GBK">
<!--网站标题-->
<title>首页--天码行空学习建站</title>
<!--引入网站的样式-->
<link type="text/css" href="css/style.css" rel="stylesheet">
<!--网站头部结束-->
<!--网站身体-->
<body bgcolor="#7fffd4">
 <div class="bgStyle">
        <div class="header">
               <div class="logo">
                      <img src="img/logo.png">
               </div>
```

```
            <div class="cssmenu">
                <ul>
                    <li><a href="index.html">首页</a></li>
                    <li><a href="about.html">企业简介</a></li>
                    <li><a href="noticeFrontList.jsp">公告</a></li>
                    <li><a href="newsFrontList.jsp">新闻</a></li>
                    <li><a href="product.html">产品介绍</a></li>
                    <li class="last"><a href="contact.html">联系我们</a></li>
                </ul>
            </div>
            <!--清除div的浮动-->
            <div class="clear"></div>
        </div>
    </div>
    <!--轮播效果-->
    <div id="fwslider" style="height: 430px;">
        <!--轮播图片-->
        <div class="slider_container">
            <div class="slide" style="opacity: 1; z-index: 1; display: block;">
                <img src="img/img2.jpg">
            </div>
        </div>
    </div>
    <div class="main_bg">
        <div class="business">业务领域 BUSINESS</div>
        <div class="wrap" style="width: 72%">
            <div class="grids">
                <div class="grid_1">
                    <a href="#" target="_blank"><img
                        src="img/pic1.png"></a>
                </div>
                <div class="grid_1">
                    <a href="#" target="_blank"><img
                        src="img/pic2.png"></a>
                </div>
                <div class="grid_1">
                    <a href="#" target="_blank"><img
                        src="img/pic3.png"></a>
                </div>
                <div class="grid_1">
                    <a href="#" target="_blank"><img
                        src="img/pic4.png"></a>
                </div>
                <div class="grid_1">
                    <a href="#" target="_blank"><img
```

```
                    src="img/pic5.png"></a>
                </div>
                <div class="clear"></div>
            </div>
        </div>
    </div>
    <div class="address" style="color: black">
        Copyright 天码行空学习建站
        <br> 天码行空学习建站 <br>
        <a href="#" style="color: black">天码行空学习建站</a>
        技术支持 <a href="../admin/login.jsp" style="color: red">  后台</a>
    </div>
</body>
</html>
```

（三）网站新闻展示功能的实现

作为企业门户网站，公司新闻是企业必不可少的一部分，在上一个任务中介绍了后台新闻管理功能的增删改查，本节实现把新闻记录展示在网站前台页面中。

首先在 News 类中，添加前台新闻列表显示方法 ListNewsFront 和新闻详细信息显示方法 FrontNewsDetail，代码如下：

```
    /**
        * 方法名：ListNewsFront
        * 功能描述：前台新闻列表页面
        * @param toPage:分页跳转页面地址，pageNum:每页显示记录数量
        * @return 数据库查询，新闻列表字符串结果集
     * Created by 天码行空
        */
    public String ListNewsFront(String toPage,String pageNum)
        {
          try
          {
                Connection Conn = DBConn.getConn();
          Statement stmt = Conn.createStatement();
          ResultSet rs = null;
                StringBuffer resultData = new StringBuffer();
                String sSql = "select * from News order by NewsID desc";
          rs = stmt.executeQuery(sSql);
          resultData=Fun.ListNewsFront(resultData,rs,toPage,pageNum);   //获取返回分页结果
                rs.close();
                stmt.close();
                Conn.close();
                return resultData.toString();
          }catch(Exception e)
```

```
            {
                return "No";
            }
        }

    /**
        * 方法名：FrontNewsDetail
        * 功能描述：前台新闻详情页面
        * @param id:文章记录唯一主键
        * @return 数据库查询，新闻详情字符串结果集
      * Created by 天码行空
        */
    public String FrontNewsDetail(String id) {
        try {
            Connection Conn = DBConn.getConn();
            Statement stmt = Conn.createStatement();
            ResultSet rs = null;
            int NewsID = Fun.StrToInt(id);
            if (NewsID == 0)
                return "No";
            else {
                try {
                    String sql = "select * from News where NewsID=" + NewsID;
                    rs = stmt.executeQuery(sql);
                    StringBuffer sb = new StringBuffer();
                    sb=Fun.FrontNewsDetail(sb, rs);
                    rs.close();
                    stmt.close();
                    Conn.close();
                    return sb.toString();
                } catch (Exception e) {
                    Conn.rollback();
                    Conn.close();
                    return "No";
                }
            }
        } catch (Exception e) {
            return "No";
        }
    }
```

在 front 目录下创建前台新闻列表显示页面 newsFrontList.jsp，其代码如下：

```
<%@ page language="java" import="java.util.*" pageEncoding="GBK"%>
<%@ page import="jspSamples.unit7.websiteSample.*"%>
<!DOCTYPE html>
<html>
```

```
<!--头部-->
<head>
<meta charset="GBK">
<!--网站标题-->
<title>新闻--天码行空学习建站</title>
<!--引入网站的样式-->
<link type="text/css" href="css/style.css" rel="stylesheet">
<!--引入网站的特效-->
<script type=" text/javascript" src="js/fwslider.js"></script>
</head>
<!--头部结束-->
<!--身体-->
<body>
 <div class="bgStyle">
      <div class="header">
           <div class="logo">
                <img src="img/logo.png">
           </div>
           <div class="pull-icon">
                <a id="pull"></a>
           </div>
           <div class="cssmenu">
                <ul>
                     <li><a href="index.html">首页</a></li>
                     <li><a href="about.html">企业简介</a></li>
                     <li><a href="noticeFrontList.jsp">公告</a></li>
                     <li><a href="newsFrontList.jsp">新闻</a></li>
                     <li><a href="product.html">产品介绍</a></li>
                     <li class="last"><a href="contact.html">联系我们</a></li>
                </ul>
           </div>
           <!--清除浮动-->
           <div class="clear"></div>
      </div>
  </div>
  <div class="second_banner">
       <img src="img/news.jpg">
  </div>

  <div class="container">
       <div class="left">
            <div class="menu_plan">
                 <div class="menu_title">
                      公司动态<br> <span>news of company</span>
                 </div>
```

```
                <ul id="tab">
                    <li><a href="#">公司新闻</a></li>
                </ul>
            </div>
        </div>
        <div class="right">
            <div class="location">
                <span>当前位置：<a href="javascript:void(0)" id="a"><a
                            href="#">公司新闻</a></a></span>
                <div class="brief" id="b">
                    <a href="#">公司新闻</a>
                </div>
            </div>
            <div style="font-size: 14px; margin-top: 53px; line-height: 36px;">
                <div id="tab_con">
                    <div id="tab_con_2"   >
                        <table style="margin-top: 70px; width: 100%">
                            <tbody>
                                <tr class="tt_bg" style="text-align:center">
                                    <td>新闻标题</td>
                                    <td>发布人</td>
                                    <td>发布时间</td>
                                    <td>详情</td>
                                </tr>
                                <%
                                    request.setCharacterEncoding("GBK");
                                    News news = new News();
                                    String pageNum = request.getParameter("intPage");
                                    String toPage = request.getContextPath() + request.
                                    getServletPath()+ "?";
                                    String sOK = news.ListNewsFront(toPage, pageNum);if (sOK.
                                    equals("No")) {
                                        out.println("数据服务器出现错误！");
                                    } else {
                                        out.println(sOK);
                                    }
                                %>
                            </tbody>
                        </table>
                    </div>

                </div>
            </div>
        </div>
    </div>
```

```
    <div class="bottom">
        <div class="footer">
            <div class="address">
                Copyright 天码行空学习建站
                <br> 天码行空学习建站 <br>
                <a href="#">天码行空学习建站</a> 技术支持 <a
                    href="../admin/login.jsp">后台</a>
            </div>
        </div>
    </div>
</body>
</html>
```

门户网站前台新闻列表页面如图 7-28 所示，此页面实现新闻列表，并显示每个新闻页面的详情链接。

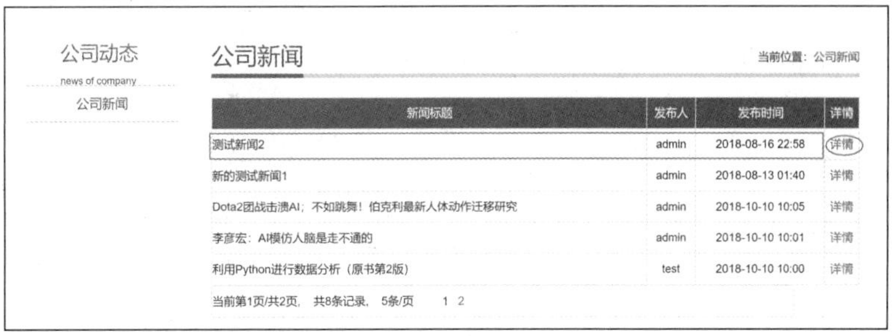

公司动态
news of company
公司新闻

公司新闻　　当前位置：公司新闻

新闻标题	发布人	发布时间	详情
测试新闻2	admin	2018-08-16 22:58	详情
新的测试新闻1	admin	2018-08-13 01:40	详情
Dota2团战击溃AI；不如跳舞！伯克利最新人体动作迁移研究	admin	2018-10-10 10:05	详情
李彦宏：AI模仿人脑是走不通的	admin	2018-10-10 10:01	详情
利用Python进行数据分析（原书第2版）	test	2018-10-10 10:00	详情

当前第1页/共2页，共8条记录，5条/页　1 2

图 7-28

接下来创建显示每条新闻详细信息的 newsFrontDetail.jsp 页面，其代码如下：

```
<%@ page language="java" import="java.util.*" pageEncoding="GBK"%>
<%@ page import="jspSamples.unit7.websiteSample.*"%>
<!DOCTYPE html>
<html>
<!--头部-->
<head>
<meta charset="GBK">
<!--网站标题-->
<title>新闻详情--天码行空学习建站</title>
<!--引入网站的样式-->
<link type="text/css" href="css/style.css" rel="stylesheet">
<!--引入网站的特效-->
<script type=" text/javascript" src="js/fwslider.js"></script>
</head>
<!--头部结束-->
<!--身体-->
<body>
 <div class="bgStyle">
     <div class="header">
```

```
            <div class="logo">
                <img src="img/logo.png">
            </div>
            <div class="pull-icon">
                <a id="pull"></a>
            </div>
            <div class="cssmenu">
                <ul>
                    <li><a href="index.html">首页</a></li>
                    <li><a href="about.html">企业简介</a></li>
                    <li><a href="noticeFrontList.jsp">公告</a></li>
                    <li><a href="newsFrontList.jsp">新闻</a></li>
                    <li><a href="product.html">产品介绍</a></li>
                    <li class="last"><a href="contact.html">联系我们</a></li>
                </ul>
            </div>
            <!--清除浮动-->
            <div class="clear"></div>
        </div>
    </div>
    <div class="second_banner">
        <img src="img/news.jpg">
    </div>
    <div class="container">
        <div class="left">
            <div class="menu_plan">
                <div class="menu_title">
                    公司动态<br> <span>news of company</span>
                </div>
                <ul id="tab">
                    <li><a href="#">公司新闻</a></li>
                </ul>
            </div>
        </div>
        <div class="right">
            <div class="location">
                <span>当前位置：<a href="javascript:void(0)" id="a"><a
                            href="#">公司新闻</a></a></span>
                <div class="brief" id="b">
                    <a href="#">公司新闻</a>
                </div>
            </div>
            <div style="font-size: 14px; margin-top: 53px; line-height: 36px;">
                <div id="tab_con">
                    <div id="tab_con_2">
```

```
                        <div class="content_main">
                            <%
                                request.setCharacterEncoding("GBK");
                                News news = new News();
                                String newsId = request.getParameter("newsId");
                                String sPage = request.getContextPath() + request.getServlet Path()+ "?";
                                        String sOK = news.FrontNewsDetail(newsId);
                                        if (sOK.equals("No")) {
                                            out.println("数据服务器出现错误！ ");
                                        } else {
                                            out.println(sOK);
                                        }
                            %>
                        </div>
                        </div>

                    </div>
                </div>
            </div>
        </div>
        <div class="bottom">
            <div class="footer">
                <div class="address">
                    Copyright 天码行空学习建站
                    <br> 天码行空学习建站 <br>
                    <a href="#">天码行空学习建站</a> 技术支持 <a
                        href="../admin/login.jsp">后台</a>
                </div>
            </div>
        </div>
    </body>
</html>
```

门户网站前台新闻的详细信息页面如图 7-29 所示。

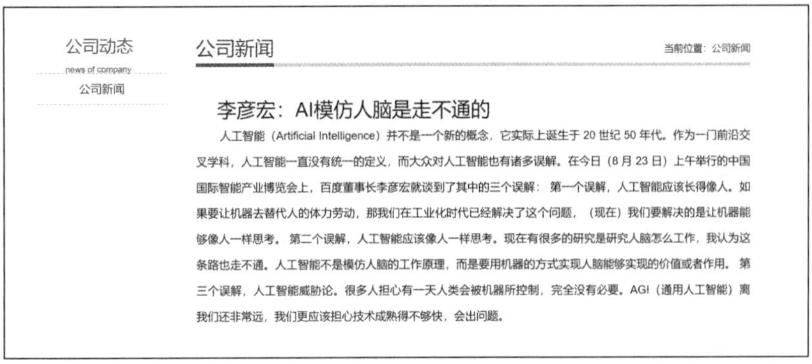

公司动态

news of company

公司新闻

公司新闻　　当前位置：公司新闻

李彦宏：AI模仿人脑是走不通的

人工智能（Artificial Intelligence）并不是一个新的概念，它实际上诞生于 20 世纪 50 年代。作为一门前沿交叉学科，人工智能一直没有统一的定义，而大众对人工智能也有诸多误解。在今日（8 月 23 日）上午举行的中国国际智能产业博览会上，百度董事长李彦宏就谈到了其中的三个误解： 第一个误解，人工智能应该长得像人。如果要让机器去替代人的体力劳动，那我们在工业化时代已经解决了这个问题，（现在）我们要解决的是让机器能够像人一样思考。 第二个误解，人工智能应该像人一样思考。现在有很多的研究是研究人脑怎么工作，我认为这条路也走不通。人工智能不是模仿人脑的工作原理，而是要用机器的方式实现人脑能够实现的价值或者作用。 第三个误解，人工智能威胁论。很多人担心有一天人类会被机器所控制，完全没有必要。AGI（通用人工智能）离我们还非常远，我们更应该担心技术成熟得不够快，会出问题。

图 7-29

提示

按照如上步骤已经实现前台门户网站的新闻展示，“公告展示”功能实现与新闻展示实现的基本方式相同，此处不再赘述，留给读者自行练习和实现，详细代码可参见本项目示例代码。已提供“天码行空”企业门户全部项目代码。

本章小结

本章通过“天码行空企业网站”介绍了完整的企业网站项目开发流程与技术实现。首先进行了系统功能分析与设计，规划了功能模块与业务流程，并且设计了数据字典；然后通过“后台管理系统主要功能”的实现，完成从项目创建到代码实现的全部流程；最后通过“企业网站主要功能实现”模块，讲解了企业网站的设计原理和与后台数据的交互显示。通过本章的完整案例，读者可以体验一个小型项目从分析设计到实现的全过程。

课后练习

编程实践题

（1）按照本章介绍的技术实现方式，实现后台管理中的“公告管理”功能模块。

（2）按照本章介绍的技术实现方式，实现前台门户网站中的“公告”展示功能。

（3）请实现“天码行空”网站的主页、企业简介、产品介绍、联系我们等静态页面。

PART08

项目八

孕婴网站的设计与实现

本章开发了一个完整的网站——孕婴网站，运用软件工程的设计思想，应用 JSP+JavaBean+Servlet 技术进行 Web 项目的实战开发。书中按照“系统功能分析→数据库设计→公共模块实现→主要页面设计→主要功能模块实现→项目运行发布”的设计与实现过程进行介绍，带领读者一步步体验项目开发的全过程。

课堂学习目标

- 系统功能分析与设计
- 系统数据库设计
- 公共模块功能的实现
- 主要页面设计与实现
- 主要功能模块的实现
- 项目运行发布

素养拓展

- 细节决定成败

素养拓展

任务一 系统功能分析与设计

任务要求

本任务要求对“孕婴网”系统功能结构、网站效果原型、系统开发环境、文件组织结构等进行分析。

任务实现

(一)系统功能结构分析

随着人们生活水平的提高，对孕产服务质量的要求越来越高，近几年提供孕婴服务的机构增多，孕婴中心展示服务优势和对外宣传的网站，已经成为大部分孕婴中心必不可少的建设内容。

本项目的“孕婴网”根据目前大部分提供孕婴服务的网站的基本需求，设计了八个功能模块。

(1)网站主页：通过主页展示整个企业门户的风格面貌与经营理念，起到提纲挈领的导航作用。

(2)关于我们：介绍会所的经营方针、优势、图片等信息。

(3)套餐活动：根据自定义的服务内容，动态显示打包服务价格。

(4)专业服务：安全措施与会所或中心可以提供的专业服务内容。

(5)企业团队：介绍团队构成，包括经理、育婴师、月嫂等团队成员。

(6)房间介绍：房间设施介绍、人员及服务介绍、房间图片展示。

(7)招贤纳士：企业岗位介绍、招聘信息发布、企业人事邮箱与电话。

(8)会员管理：会员注册、会员登录、会员基本信息修改等。

系统功能结构如图 8-1 所示。

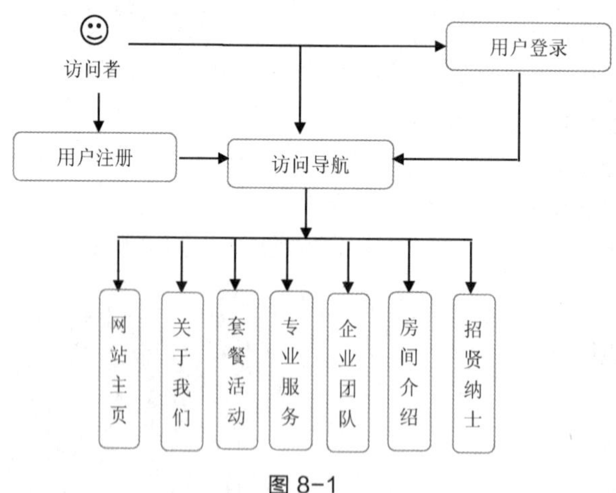

图 8-1

(二)网站效果原型

为了让读者对本网站有个初步的了解与认识，下面给出本网站的几个主要页面的效果图。

网站主页如图 8-2 所示；会员注册如图 8-3 所示；套餐活动如图 8-4 所示；专业服务如图 8-5

所示；企业团队如图 8-6 所示；房间介绍如图 8-7 所示。

图 8-2

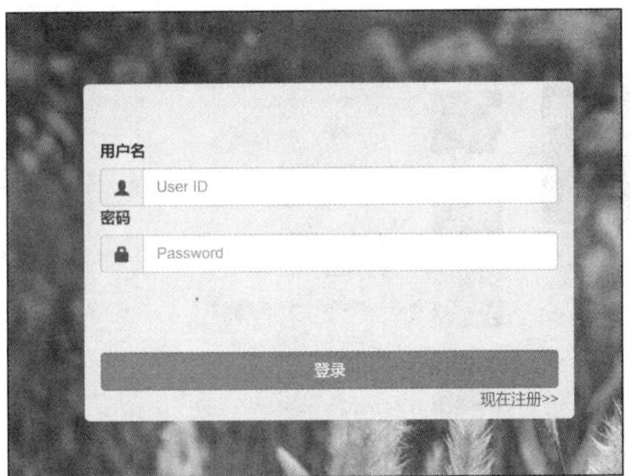

图 8-3

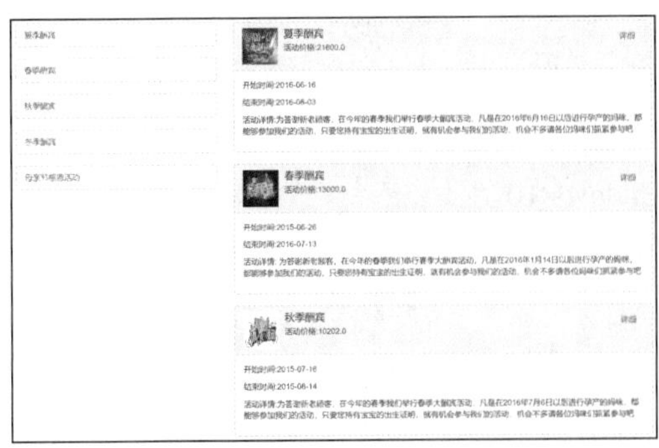

图 8-4

图 8-5

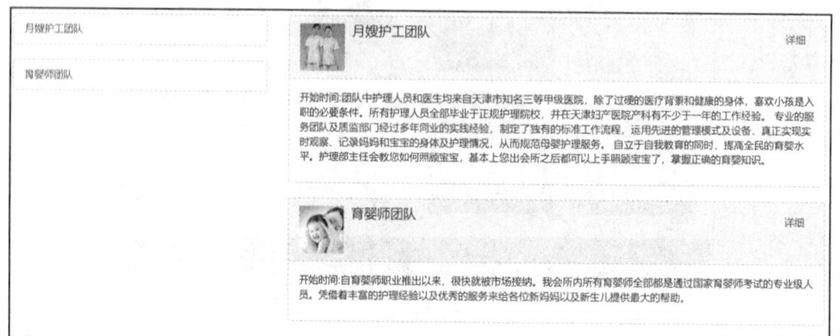

图 8-6

图 8-7

（三）系统开发环境

本系统的软件开发及运行环境如下。

➢ 操作系统：Windows 7，Windows 8.1，Windows 10。
➢ JDK 环境：Java SE Development Kit (JDK) version 8。
➢ 开发工具： IntelliJ IDEA 2018。
➢ Web 服务器：Tomcat 8.0。
➢ 数据库：MySQL-5.7（配置版）。
➢ 浏览器：推荐谷歌浏览器或者火狐浏览器。

➢ 分辨率：推荐分辨率为 1024 像素 × 768 像素。

（四）文件组织结构

在进行“孕婴网”开发之前，要对网站整体组织架构进行规划，对网站中使用的文件进行合理分类，分别放置于不同的文件夹下。通过对文件夹组织结构的规划，可以确保网站文件的目录明确、条理清晰，便于网站的更新和维护。本网站的文件夹组织规划如图 8-8 所示。

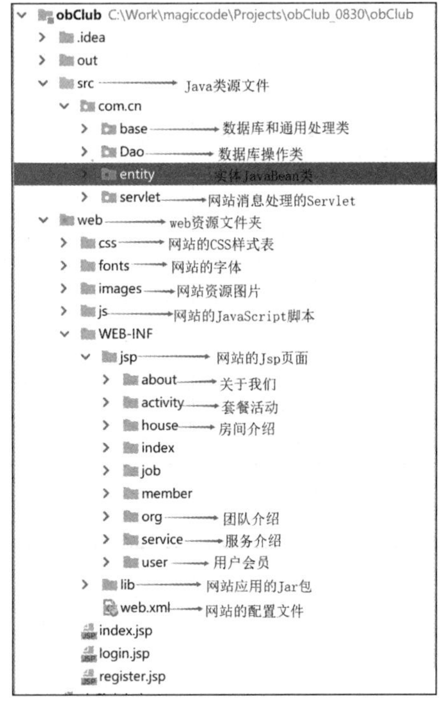

图 8-8

任务二　系统数据库设计

任务要求

本任务要求通过对“孕婴网”的功能分析，进行数据库设计，此处使用的数据库工具为 MySQL-5.7（配置版），读者可以根据需要选择相应的数据库产品。

任务实现

（一）数据库设计

根据系统功能分析，要对会员与用户、套餐及活动、企业团队信息、房型介绍、专业服务、招聘信息等模块进行信息发布与展示，所以在此设计至少 7 个表以支撑目前的系统功能，此处设计的数据库名称为 obclub。

（二）数据表设计

➢ 会员信息表（MEMBER_INFO）：主要用来保存系统的会员信息，如表 8-1 所示。

表 8-1 会员信息表

字段名	数据类型	是否 Null 值	默认值	描述
ID	BIGINT	否		主键，自增
MEM_USERNAME	VARCHAR(30)	否		会员用户名
MEM_PASSWORD	VARCHAR(30)	否		会员密码
LEVEL_ID	BIGINT	否		
MEM_NAME	VARCHAR(16)	否		会员姓名
MEM_SEX	CHAR(2)	是	男	性别（男、女）
MEM_ADDRESS	VARCHAR(200)	是		家庭住址
MEM_TEL	VARCHAR(32)	是		联系电话
MEM_EMAIL	VARCHAR(64)	是		联系邮箱
REG_TIME	DATETIME	否		注册日期
CARD_NO	VARCHAR(32)	是		会员卡号
STATUS	VARCHAR(32)	是		会员状态
MEM_SCORE	FLOAT	是		会员积分
MEM_PIC	BLOB	是		会员头像

➢ 套餐及活动表（ACTIVITY_INFO）：主要用来保存孕婴网的优惠套餐和活动信息，如表 8-2 所示。

表 8-2 套餐及活动表

字段名	数据类型	是否 Null 值	默认值	描述
ID	BIGINT	否		主键，自增
ACT_TITLE	VARCHAR(128)	否		套餐活动标题
ACT_EXTRA	VARCHAR(128)	是		套餐活动附加标题
ACT_DETAIL	VARCHAR(1024)	是		套餐活动描述
ACT_PIC	BLOB	是		套餐活动图片
ACT_PRICE	FLOAT	是		套餐活动价格
START_DATE	DATETIME	是		起始日期
END_DATE	DATETIME	是		截止日期
CREATEDATE	DATETIME	否	Getdate()	记录创建时间
SHOWORDER	INT	否		显示顺序
IFSHOW	CHAR(2)	否	是	是否显示在网站

➢ 企业团队信息表（ORG_INFO）：主要用来保存企业具备的月嫂和育英团队信息，如表 8-3 所示。

表 8-3　企业团队信息表

字段名	数据类型	是否 Null 值	默认值	描述
ID	BIGINT	否		主键，自增
ORG_TITLE	VARCHAR(64)	否		团队信息标题
ORGB_DETAIL	VARCHAR(512)	是		团队信息介绍
ORG_PIC	BLOB	是		团队信息相关图片
CREATEDATE	DATETIME	否	Getdate()	记录创建时间
SHOWORDER	INT	否		显示顺序
IFSHOW	CHAR(2)	否	是	是否显示在网站

➢ 房型信息表（HOUSE_INFO）：主要用来保存孕婴中心提供的房型和价格信息，如表 8-4 所示。

表 8-4　房型信息表

字段名	数据类型	是否 Null 值	默认值	描述
ID	BIGINT	否		主键，自增
HOUSE_TITLE	VARCHAR(64)	否		房型名称
HOUSE_PRICE	FLOAT	是		房型价格
HOUSE_DETAIL	VARCHAR(512)	是		房型描述
HOUSE_PIC	BLOB	是		房型相关图片
CREATEDATE	DATETIME	否	Getdate()	记录创建时间
SHOWORDER	INT	否		显示顺序
IFSHOW	CHAR(2)	否	是	是否显示在网站

➢ 专业服务类型表（SERVICE_TYPE）：主要用来保存孕婴中心可以提供的专业服务类型，如表 8-5 所示。

表 8-5　专业服务类型表

字段名	数据类型	是否 Null 值	默认值	描述
ID	BIGINT	否		主键，自增
TYPE_TITLE	VARCHAR(64)	否		类型名称
CREATEDATE	DATETIME	否	Getdate()	记录创建时间
SHOWORDER	INT	否		显示顺序
IFSHOW	CHAR(2)	否	是	是否显示在网站

➢ 专业服务信息表（SERVICE_INFO）：主要用来保存孕婴中心可以提供的专业服务信息，如表 8-6 所示。

表 8-6　专业服务信息表

字段名	数据类型	是否 Null 值	默认值	描述
ID	BIGINT	否		主键，自增
TYPE_ID	BIGINT	否		外键（所属服务类型）
SER_TITLE	VARCHAR(64)	否		服务名称

续表

字段名	数据类型	是否 Null 值	默认值	描述
SER_DETAIL	VARCHAR(512)	是		服务描述
SER_PIC	BLOB	是		服务相关图片
CREATEDATE	DATETIME	否	Getdate()	记录创建时间
SHOWORDER	INT	否		显示顺序
IFSHOW	CHAR(2)	否	是	是否显示在网站

➢ 招贤纳士信息表（JOB_INFO）：主要用来保存孕婴网招聘信息，如表 8-7 所示。

表 8-7　招贤纳士信息表

字段名	数据类型	是否 Null 值	默认值	描述
ID	BIGINT	否		主键，自增
JOB_TITLE	VARCHAR(64)	否		职位名称
JOB_DETAIL	VARCHAR(512)	是		职位描述
CREATEDATE	DATETIME	否	Getdate()	记录创建时间
SHOWORDER	INT	否		显示顺序
IFSHOW	CHAR(2)	否	是	是否显示在网站

读者可以根据自己理解的孕婴网站功能进行网站功能的设计与扩展，并进行表结构的修改与扩展。

任务三　公共模块功能的实现

任务要求

本任务要求实现网站的公共功能模块。在开发过程中经常会用到一些公共模块，例如，数据库连接及操作的类、保存分页的类、将文件转换为二进制或者日期处理的工具类、与数据表直接关联的实体类等。在开发网站时，要先设计与实现这些公共模块。

任务实现

（一）创建项目

微课：创建“孕婴网”网站项目

本章使用 IntelliJ IDEA 2018 集成开发环境（以下简称 IDEA）实现项目的建立，首先创建“孕婴网”网站项目，具体步骤如下：

打开 IDEA 后，依次选择 File->New->Project 菜单项，如图 8-9 所示。

单击 Project 之后，出现 New Project 对话框，左侧区域中选择 Java 选项，右侧区域中选择 Java EE->Web Application 选项，然后单击 Next 按钮，如图 8-10 所示。

提示

在 Project SDK 处，请选择本机安装的 JDK 1.8 版本。如果未显示本机安装的 JDK，请查阅 IDEA 如何配置 JDK 的相关资料。

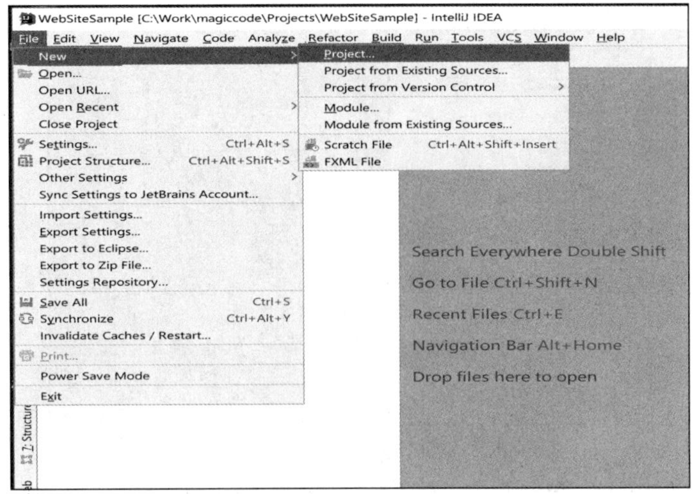

图 8-9

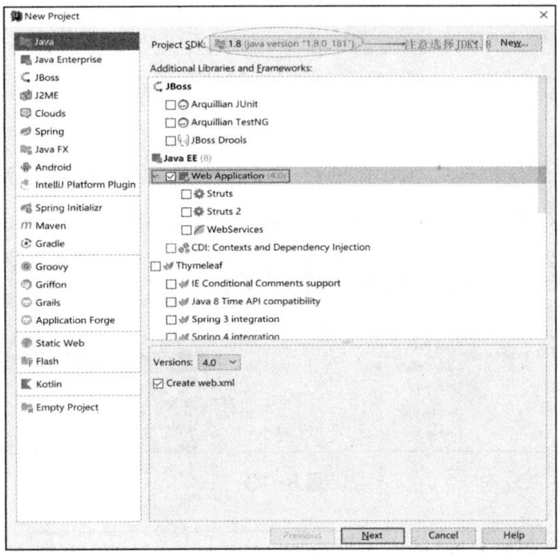

图 8-10

选择所创建项目类型之后，请为项目命名，此处将 Project name 命名为 obclub（要符合文件名命名规范），并且选择项目的存储路径，然后单击 Finish 按钮，如图 8-11 所示。

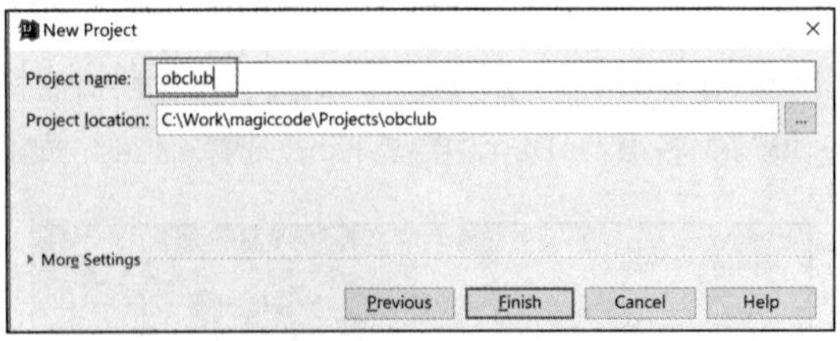

图 8-11

创建 obclub 项目以后，显示了项目的整体结构，一般情况下，把包和类的源码放在 src 目录下，把网站的静态与动态网页及网站相关资源放在 web 目录下，如图 8-12 所示。

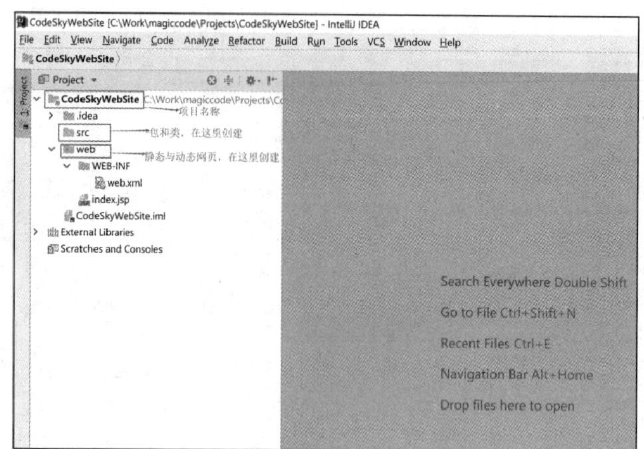

图 8-12

（二）数据库连接类的实现

数据库连接类在所有数据库访问的过程中都会用到，所以把数据库连接类提炼出来作为公共模块中的一个类，主要用到的方法是建立数据库连接 getConn。根据功能要求，数据库连接类 BaseDao 实现的具体步骤如下。

首先建立整个项目源文件的包结构，在创建类之前先创建包。右键单击 src->New->Package，如图 8-13 所示；在输入包名处命名包为 com.cn.base，单击 OK 按钮，如图 8-14 所示。

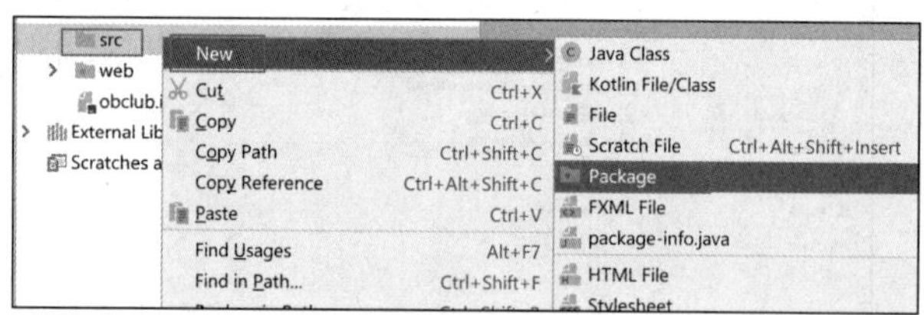

图 8-13

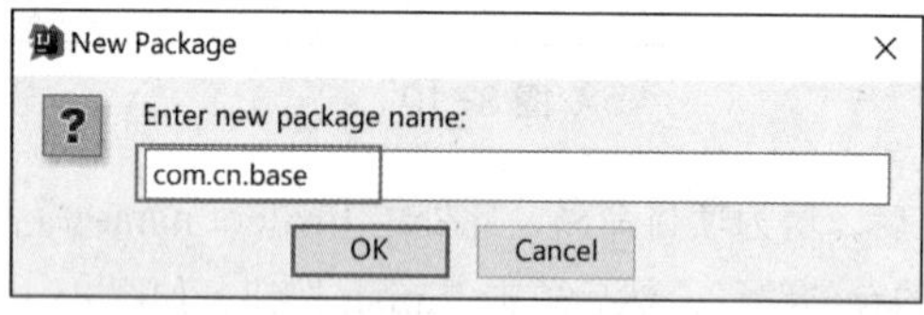

图 8-14

创建数据库访问 BaseDao 类，如图 8-15 所示，在刚创建的包处单击右键选择 New->Java Class，出现创建 Class 的对话框，填写类名 BaseDao，类型 Kind 选择 Class，如图 8-16 所示。

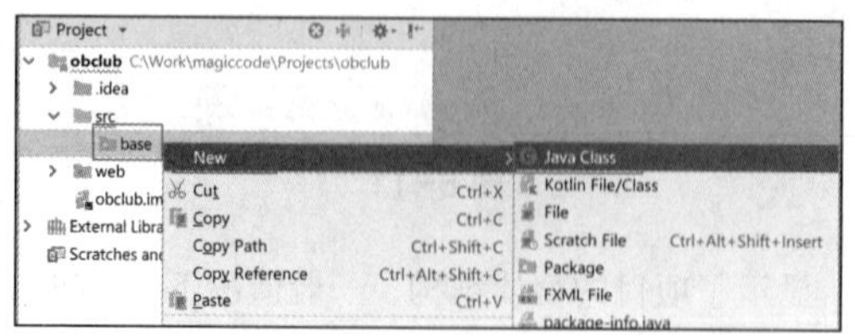

图 8-15

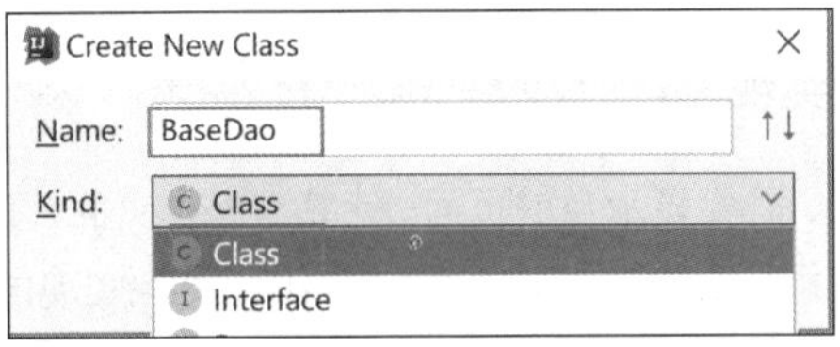

图 8-16

数据库访问 BaseDao 类主要实现数据库访问与连接，具体实现代码如下：

```
package com.cn.base;

import com.mysql.jdbc.Connection;

import java.sql.DriverManager;
import java.sql.SQLException;

public class BaseDao {

    public static Connection getConn() {
        String driver = "com.mysql.jdbc.Driver";
        String url = "jdbc:mysql://127.0.0.1:3306/obClub?useUnicode=true&characterEncoding=GBK";
        String username = "root";
        String password = "";
        Connection conn = null;
        try {
            Class.forName(driver); //classLoader,加载对应驱动
            conn = (Connection) DriverManager.getConnection(url, username, password);
        } catch (ClassNotFoundException e) {
            e.printStackTrace();
         } catch (SQLException e) {
            e.printStackTrace();
         }
         return conn;
    }
}
```

getConn 方法中 mysql 的数据库驱动设定为：driver = " com.mysql.jdbc.Driver ";，数据库访问的地址设定为：url = " jdbc:mysql://127.0.0.1:3306/obClub?useUnicode=true&characterEncoding=GBK ";，此 url 中有 mysql 服务器的地址，数据库名称 obClub，访问 mysql 的用户名 root，访问 mysql 数据库的密码为空。

数据库驱动类设定为 MySqlDriver = " com.mysql.jdbc.Driver "，此驱动类的压缩包为 mysql-connector-java-5.1.34-bin.jar，需要将本书提供的项目资源文件中的此压缩包复制到 “web\WEB-INF\lib” 目录下。

（三）保存分页类的实现

此处实现的是分页工具类，需要显示当前页面、计算与保存页面、取得页面列表、取得当前页面、取得所有页面等。根据功能要求，分页工具泛型类 PageUtil<T>实现的具体步骤如下：

首先建立保存页面的实体类 Page，在包 com.cn.entity 下建立 Page 类，具体代码如下：

```
package com.cn.entity;

public class Page {

    private int pageCurrent = 1;            //当前页
    private int pageSize = 3;               //一页显示数目
    private int pageTotal = 1;              //  显示多少页

    public Page() {
    }

    public int getPageCurrent() {
        return pageCurrent;
    }

    public void setPageCurrent(int pageCurrent) {
        this.pageCurrent = pageCurrent;
    }

    public int getPageSize() {
        return pageSize;
    }

    public void setPageSize(int pageSize) {
        this.pageSize = pageSize;
    }

    public int getPageTotal() {
        return pageTotal;
    }

    public void setPageTotal(int pageTotal) {
        this.pageTotal = pageTotal;
    }
}
```

在公共模块包 com.cn.base 下建立分页工具类 PageUtil<T>，具体代码如下：

```
package com.cn.base;

import com.cn.entity.Page;
```

```
import javax.servlet.http.HttpServletRequest;
import java.util.ArrayList;
import java.util.List;

public class PageUtil<T> {
    /**
     * 计算总页数
     * @param pageSize    一页显示的内容数量
     * @param listSize
     * @return
     */
    private int showPage(int pageSize, int listSize){
        return (listSize % pageSize == 0) ? (listSize / pageSize) : (listSize / pageSize + 1);
    }

    /**
     * 每页显示第一条对应list的索引
     * @param pageSize
     * @param pageCurrent
     * @return
     */
    private int pageStart(int pageSize, int pageCurrent){
        return pageCurrent * pageSize - pageSize;
    }

    /**
     * 获取当前页list
     * @param page
     * @param list
     * @return
     */
    public List<T> getList(Page page, List<T> list){
        List<T> arrayList = new ArrayList<>();

        if(page.getPageCurrent() > 0) {
            int pageStart = pageStart(page.getPageSize(), page.getPageCurrent());
// 从list第几条开始

            for (int i = page.getPageSize(); i > 0; i--, pageStart++) {
                if ((pageStart) < list.size()) {
                    arrayList.add(list.get(pageStart));
                }
            }
            return arrayList;
```

```
            }
            return null;
        }

        /**
         * 获取当前页
         * @param request
         * @return
         */
        private int getPageCurrent(HttpServletRequest request){
            String pageCurrent = request.getParameter("pageCurrent");
            if (pageCurrent != null){
              return Integer.parseInt(request.getParameter("pageCurrent"));        //  当前页
            }
            return 1;
        }

        /**
         * 获取Page所有属性
         * @param list      右侧显示list
         * @param pageSize      一页显示内容数量
         * @param request
         * @return
         */
        public Page allPage(List<T> list, int pageSize, HttpServletRequest request){
              Page page = new Page();
              page.setPageSize(pageSize);
              page.setPageTotal(showPage(pageSize, list.size()));
              page.setPageCurrent(getPageCurrent(request));        //  当前页
              return page;
        }
}
```

（四）基本工具类的实现

每个项目都有自己的基本工具类，基本工具类作为公共模块功能类，分类特性为与项目本身的业务无关或者相关性小，最常用的一些基本操作的功能集合包括处理日期、处理图片、处理二进制、处理加密解密、处理缓存等。根据功能要求，此项目中基本工具类 Util<T>实现的具体操作是在包 com.cn.base 下建立 Util<T>类，具体代码如下：

```
package com.cn.base;

import org.apache.commons.beanutils.BeanUtils;

import javax.servlet.http.HttpServletRequest;
import java.io.ByteArrayOutputStream;
```

```
import java.io.File;
import java.io.FileInputStream;
import java.io.OutputStream;
import java.text.SimpleDateFormat;
import java.util.ArrayList;
import java.util.Date;
import java.util.Enumeration;
import java.util.List;
import java.util.UUID;

public class Util<T> {

    /**
     * 将文件转换为二进制
     * @param file
     * @return
     * @throws Exception
     */
    public static byte[] file2byte(File file) throws Exception {
        byte[] buffer = new byte[(int)file.length()];

        FileInputStream fis = new FileInputStream(file);
        ByteArrayOutputStream bos = new ByteArrayOutputStream();

        byte[] b = new byte[4096];
        int n;
        while ((n = fis.read(b)) != -1) {
            bos.write(b, 0, n);

        }
        fis.close();
        bos.close();
        buffer = bos.toByteArray();
        return buffer;
    }
    /**
     * 改为时间戳格式
     * @param date
     * @return
     */
    public static String changeDate(Date date){
        SimpleDateFormat formatter = new SimpleDateFormat("yyyy-MM-dd HH:mm:ss");
        String dateString = formatter.format(date);
        return dateString;
    }
```

```
}
```

（五）实体类的实现

实体类就是由属性及所对应的 Getter()和 Setter()方法组成的类，实体类通常与数据表关联，作为页面和类传递每个实体表数据的容器。在本孕婴网中涉及较多的表，需要用实体类关联和表现，在本节中为了阐述清晰，只介绍有代表性的“房型介绍”主要实体类，其余雷同，不再赘述，读者可自行完成代码或者参考本章提供的完整“孕婴网”项目代码的实体类部分。根据功能要求，此项目中“房型介绍”主要实体类的实现具体步骤如下：

在 com.cn.entity 包下，创建 HouseInfo 类，用于存储房型信息（具体属性的含义，请参考表 8-4 的 HouseInfo 表的数据字典说明），具体代码如下：

```
package com.cn.entity;

import java.util.Arrays;
import java.util.Date;

public class HouseInfo {
    private Long id;
    private String houseTitle;
    private Float housePrice;
    private String houseDetail;
    private byte[] housePic;
    private Date createdate;
    private Integer showorder;
    private String ifshow;

    public HouseInfo() {
    }
    public HouseInfo(String houseTitle, Date createdate, Integer showorder,
                String ifshow) {
        this.houseTitle = houseTitle;
        this.createdate = createdate;
        this.showorder = showorder;
        this.ifshow = ifshow;
    }
    public HouseInfo(String houseTitle, Float housePrice, String houseDetail,
                byte[] housePic, Date createdate, Integer showorder, String ifshow) {
        this.houseTitle = houseTitle;
        this.housePrice = housePrice;
        this.houseDetail = houseDetail;
        this.housePic = housePic;
        this.createdate = createdate;
        this.showorder = showorder;
        this.ifshow = ifshow;
```

```
    }
    public Long getId() {
        return id;
    }
    public void setId(Long id) {
        this.id = id;
    }
    public String getHouseTitle() {
        return houseTitle;
    }
    public void setHouseTitle(String houseTitle) {
        this.houseTitle = houseTitle;
    }
    public Float getHousePrice() {
        return housePrice;
    }
    public void setHousePrice(Float housePrice) {
        this.housePrice = housePrice;
    }
    public String getHouseDetail() {
        return houseDetail;
    }
    public void setHouseDetail(String houseDetail) {
        this.houseDetail = houseDetail;
    }
    public byte[] getHousePic() {
        return housePic;
    }
    public void setHousePic(byte[] housePic) {
        this.housePic = housePic;
    }
    public Date getCreatedate() {
        return createdate;
    }
    public void setCreatedate(Date createdate) {
        this.createdate = createdate;
    }
    public Integer getShoworder() {
        return showorder;
    }
    public void setShoworder(Integer showorder) {
        this.showorder = showorder;
    }
    public String getIfshow() {
        return ifshow;
```

```
        }
        public void setIfshow(String ifshow) {
            this.ifshow = ifshow;
        }
        @Override
        public String toString() {
            return "HouseInfo{" +
                    "id=" + id +
                    ", houseTitle='" + houseTitle + '\'' +
                    ", housePrice=" + housePrice +
                    ", houseDetail='" + houseDetail + '\'' +
                    ", housePic=" + Arrays.toString(housePic) +
                    ", createdate=" + createdate +
                    ", showorder=" + showorder +
                    ", ifshow='" + ifshow + '\'' +
                    '}';
        }
    }
```

任务四　主要页面设计与实现

任务要求

本任务要求设计与实现“孕婴网”整体布局和主要页面静态代码部分。此任务中介绍的主要页面为：网站主页、会员登录页、房型展示页。由于整个网站的页面较多，技术雷同，其他页面不再赘述，读者可自行完成。

任务实现

（一）网站主页界面设计

当用户访问“孕婴网”时，首先进入的是网站的主页面。“孕婴网”的主界面主要包括以下四个部分。

（1）Banner 信息栏：用于显示网站的 Logo 和会员登录、注册、退出入口。

（2）导航栏：主要用于显示网站的导航信息，提供到各个版块的一级页面的链接。

（3）主显示区：主要显示“房间介绍”“主页服务”“套餐及活动”版块的定制内容。

（4）版权信息：主要用于显示版权和联系方式。

下面显示本网站中设计的主要界面，如图 8-17 所示。

主页采用 DIV+CSS 页面布局。本网站使用的是 Bootstrap 前端架构进行布局，使用 Bootstrap 提供的相关 css 样式，读者可根据个人掌握的技术进行页面布局实现。首先创建网站的资源（css、fonts、images、js）文件夹，并引入相关资源文件（按照提供的相关网站项目文件，可以直接把资源文件拷贝到项目下）。资源文件的主要目录结构如图 8-18 所示。

图 8-17

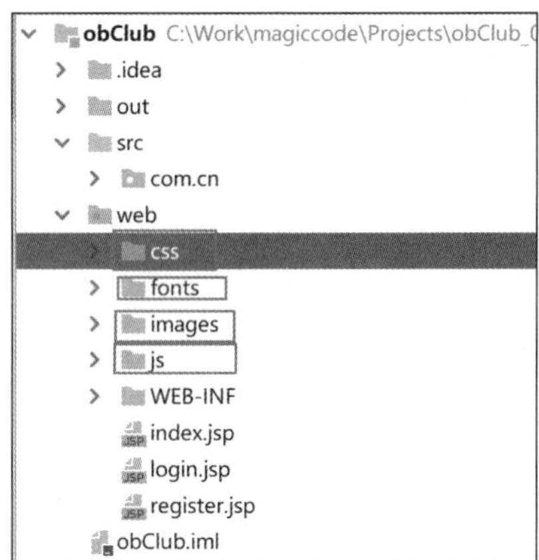

图 8-18

页面导航部分的主要布局代码如下：

```
<div class="nav">
   <a style="text-decoration: none" href="#">
     <div>
       <img width="50" class="logo" src="/images/logo.png">
         
       <img class="banner" src="/images/banner.png">
     </div>
   </a>
   <div class="tag">
```

```
        <div>
            <span class="glyphicon glyphicon-phone">联系电话</span>
        </div>
        <div id="login-btn">
            <span class="glyphicon glyphicon-user" id="span">admin01</span>
        </div>
        <div id="quit">
            <span id="quit-span">退出</span>
        </div>
    </div>
</div>
```

页面轮播图部分的主要代码如下：

```
<div class="slider row">
    <div class="carousel slide" id="slider" ng-controller="sliderCtrl" data-ride="carousel">
        <ol class="carousel-indicators">
            <li data-slide-to="0" data-target="#slider"></li>
            <li data-slide-to="1" data-target="#slider"></li>
            <li data-slide-to="2" data-target="#slider"></li>
            <li class="active" data-slide-to="3" data-target="#slider"></li>
        </ol>
        <div class="carousel-inner" role="listbox">
            <div class="item">
                <img width="100%" height="400" alt="" src="images/slider2.png">
            </div>
            <div class="item">
                <img width="100%" height="400" alt="" src="images/slider4.png">
            </div>
            <div class="item">
                <img width="100%" height="400" alt="" src="images/slider5.png">
            </div>
            <div class="item active">
                <img width="100%" height="400" alt="" src="images/slider3.png">
            </div>
        </div>
        <a class="left carousel-control" role="button" href=".slide" data-slide="prev"> <span class=
        "glyphicon glyphicon-chevron-left" aria-hidden="true"></span> <span class="sr-only">Previous
        </span> </a>
        <a class="right carousel-control" role="button" href=".slide" data-slide="next"><span class=
        "glyphicon glyphicon-chevron-right" aria-hidden="true"></span> <span class="sr-only">next
        </span></a>
    </div>
</div>
```

页面主显示区的主要布局代码如下：

```
<div class="container">
    <div class="more-list">
        <div class="more-list-header">
```

```
        <h4>房间介绍</h4> <small>room</small>
        <a href="house.jsp">更多&gt;&gt;</a>
    </div>
    <div class="content" style="margin-top: 3px">
        <div class="row">
            <div class="col-lg-3">
                <p>古风典范</p>
            </div>
            </div>
            <div class="col-lg-3">
                <div class="img-thumbnail" style="float: left">
                    <p>经济普居型</p>
                </div>
            </div>
            <div class="col-lg-3">
                <div class="img-thumbnail" style="float: left">
                    <p>温馨家居</p>
                </div>
            </div>
            <div class="col-lg-3">
                <div class="img-thumbnail" style="float: left">
                    <p>淡粉色英伦式</p>
                </div>
            </div>
        </div>
    </div>
</div>
<div class="more-list">
    <div class="more-list-header">
        <h4>专业服务</h4><small>server</small>
        <a href="service.jsp">更多&gt;&gt;</a>
    </div>
    <div class="content" style="margin-top: 3px">
        <div class="row">
            <div class="col-lg-3">
                <div class="img-thumbnail">
                    <p>全程心理辅导</p>
                </div>
            </div>
            <div class="col-lg-3">
                <div class="img-thumbnail">
                    <p>瑜伽健身</p>
                </div>
            </div>
            <div class="col-lg-3">
```

```
                    <div class="img-thumbnail">
                        <p>中医养生调理</p>
                    </div>
                </div>
                <div class="col-lg-3">
                    <div class="img-thumbnail">
                        <p>月嫂推荐</p>
                    </div>
                </div>
            </div>
        </div>
    </div>
    <div class="more-list">
        <div class="more-list-header">
            <h4>套餐及活动</h4> <small>active</small>
        </div>
        <div class="content" style="margin-top: 3px">
            <div class="row">
                <div class="col-lg-3">
                    <div class="img-thumbnail" style="float: left">
                        <p>夏季酬宾</p>
                    </div>
                </div>
                <div class="col-lg-3">
                    <div class="img-thumbnail" style="float: left">
                        <p>春季酬宾</p>
                    </div>
                </div>
                <div class="col-lg-3">
                    <div class="img-thumbnail" style="float: left">
                        <p>秋季酬宾</p>
                    </div>
                </div>
                <div class="col-lg-3">
                    <div class="img-thumbnail" style="float: left">
                        <p>冬季酬宾</p>
                    </div>
                </div>
            </div>
        </div>
    </div>
</div>
```

版权信息（页脚）部分的主要代码如下：

```
<div class="foot">
    <span>联系电话：15111111111 邮箱：obClub@qq.com</span>    
```

```
        <span>详细地址：孕婴中心地址</span>
    </div>
```

（二）会员登录页面设计

当会员进入网站后，可以单击顶部的会员登录图标进行会员登录，输入用户名和密码后，单击“登录”按钮进行验证，页面如图 8-19 所示。

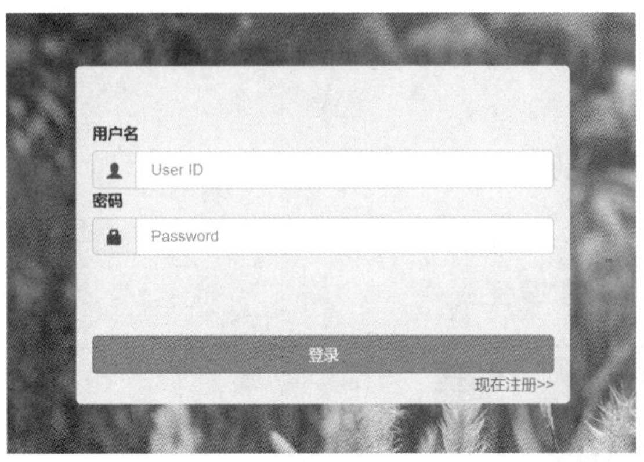

图 8-19

会员登录页面的布局实现代码如下：

```
<div class="login">
        <div class="col-lg-offset-4 col-md-offset-4 col-sm-offset-4 col-lg-4 col-md-4 col-sm-4"
        style="height: 300px;   border-radius: 5px; background-color: #EFEFEF; padding-top: 50px;
        padding- bottom: 15px">
            <form action="login" method="post">
                <div class="form-list">
                    <label class="control-label" for="txt_userId">用户名</label>
                    <div class="input-group">
                        <span class="input-group-addon">
                            <span class="glyphicon glyphicon-user"></span>
                        </span>
                        <input name="username" class="form-control" style="background-color: white"
                        required="" type="text" size="16" placeholder="User ID" value="">
                    </div>
                </div>
                <div class="form-group">
                    <label class="control-label" for="txt_password">密码</label>
                    <div class="input-group">
                        <span class="input-group-addon">
                            <span class="glyphicon glyphicon-lock"></span>
                        </span>
                        <input name="password" class="form-control" style="background-color: white"
                        required="" type="password" size="16" placeholder="Password" value="">
                    </div>
```

```
            </div>
            <div class="form-group" style="margin-top: 30px">
                <input name="submit" class="form-control btn btn-success" type="submit" value="登录">
                <a style="float: right" href="register.jsp">现在注册&gt;&gt;</a>
            </div>

        </form>

    </div>
</div>
```

（三）房型展示页面设计

房型展示页面是典型的一个列表式二级页面，使用缩略图和列表显示房型的简介信息，页面如图 8-20 所示。

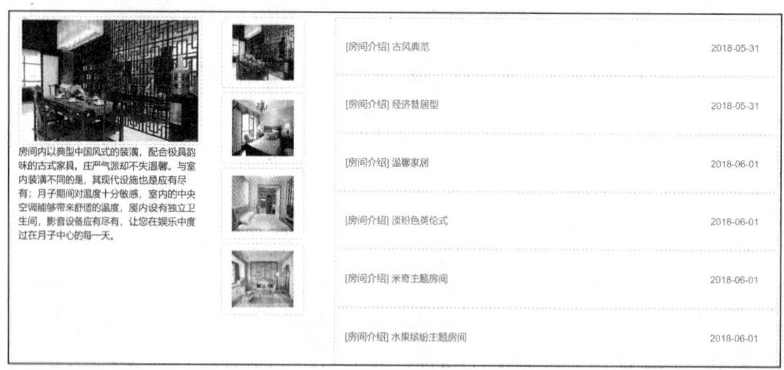

图 8-20

房型展示页面的主要布局实现代码如下：

```
<div class="container">
    <div class="room" ng-controller="roomCtrl" ng-init="getList()">
        <img style="margin-bottom: 10px" src="/images/slider2.png">
        <div class="row">
            <div class="content col-lg-5">
                <div class="col-xs-8">
                    <div ng-if="active">
                        <img height="209" class="img-thumbnail" id="house_pic" alt="" src="/images/
                        houseShowPic?id=9">
                        <p id="house_detail" ng-bind="active.houseTitle">房间内以典型中国风式的装潢，配
                        合极具韵味的古式家具。庄严气派却不失温馨。与室内装潢不同的是，其现代设施也是
                        应有尽有；月子期间对温度十分敏感，室内的中央空调能够带来舒适的温度，屋内设有
                        独立卫生间，影音设备应有尽有，让您在娱乐中度过在月子中心的每一天。</p>
                    </div>
            </div>
            <div class="col-xs-4">
                        <ul class="list-group" style="margin-bottom:6px">
```

```
                <li class="list-group-item" ng-class="{\'list-group-item-info\':isActive (item)}"
                ng-repeat="item in data | limitTo:4">
                    <a onclick="mOut(0)"><img class="img-responsive" style="height: 80px" alt=""
                    src="/images/houseShowPic?id=9"></a>
                </li>
            </ul>
            <ul class="list-group" style="margin-bottom:6px">
                <li class="list-group-item" ng-class="{\'list-group-item-info\':isActive (item)}"
                ng-repeat="item in data | limitTo:4">
                    <a onclick="mOut(1)"><img class="img-responsive" style="height: 80px" alt=""
                    src="/images/houseShowPic?id=11"></a>
                </li>
            </ul>
            <ul class="list-group" style="margin-bottom:6px">
                <li class="list-group-item" ng-class="{\'list-group-item-info\':isActive(item)}"
                ng-repeat="item in data | limitTo:4">
                    <a onclick="mOut(2)"><img class="img-responsive" style="height: 80px" alt=""
                    src="/images/houseShowPic?id=12"></a>
                </li>
            </ul>
            <ul class="list-group" style="margin-bottom:6px">
                <li class="list-group-item" ng-class="{\'list-group-item-info\':isActive(item)}"
                ng-repeat="item in data | limitTo:4">
                    <a onclick="mOut(3)"><img class="img-responsive" style="height: 80px" alt=""
                    src="/images/houseShowPic?id=13"></a>
                </li>
            </ul>
    </div>
</div>
<div class="content col-lg-7">
    <ul class="list-group" style="margin-bottom:1px">
        <li class="list-group-item">
            <a class="row" href="house_detail.jsp?id=9">
                <span class="pull-left targe">[房间介绍] 古风典范</span>
                <span class="pull-right date">2018-05-31   </span>
            </a>
        </li>
    </ul>
    <ul class="list-group" style="margin-bottom:1px">
        <li class="list-group-item">
            <a class="row" href="house_detail.jsp?id=11">
                <span class="pull-left targe">[房间介绍] 经济普居型</span>
                <span class="pull-right date">2018-05-31   </span>
            </a>
        </li>
    </ul>
```

```
<ul class="list-group" style="margin-bottom:1px">
    <li class="list-group-item">
        <a class="row" href="house_detail.jsp?id=12">
            <span class="pull-left targe">[房间介绍] 温馨家居</span>
            <span class="pull-right date">2018-06-01   </span>
        </a>
    </li>
</ul>
<ul class="list-group" style="margin-bottom:1px">
    <li class="list-group-item">
        <a class="row" href="house_detail.jsp?id=13">
            <span class="pull-left targe">[房间介绍] 淡粉色英伦式</span>
            <span class="pull-right date">2018-06-01   </span>
        </a>
    </li>
</ul>
<ul class="list-group" style="margin-bottom:1px">
    <li class="list-group-item">
        <a class="row" href="house_detail.jsp?id=14">
            <span class="pull-left targe">[房间介绍] 米奇主题房间</span>
            <span class="pull-right date">2018-06-01   </span>
        </a>
    </li>
</ul>
<ul class="list-group" style="margin-bottom:1px">
    <li class="list-group-item">
        <a class="row" href="house_detail.jsp?id=15">
            <span class="pull-left targe">[房间介绍] 水果缤纷主题房间</span>
            <span class="pull-right date">2018-06-01   </span>
        </a>
    </li>
</ul>
<ul class="list-group" style="margin-bottom:1px">
    <li class="list-group-item">
        <a class="row" href="house_detail.jsp?id=16">
            <span class="pull-left targe">[房间介绍] HelloKity粉色主题房间</span>
            <span class="pull-right date">2018-06-01   </span>
        </a>
    </li>
</ul>
<ul class="list-group" style="margin-bottom:1px">
    <li class="list-group-item">
        <a class="row" href="house_detail.jsp?id=17">
            <span class="pull-left targe">[房间介绍] 欧式风情</span>
            <span class="pull-right date">2018-06-01   </span>
```

```
                    </a>
                </li>
            </ul>
            <ul class="list-group" style="margin-bottom:1px">
                <li class="list-group-item">
                    <a class="row" href="house_detail.jsp?id=18">
                        <span class="pull-left targe">[房间介绍]  阳光美利坚</span>
                        <span class="pull-right date">2018-06-01   </span>
                    </a>
                </li>
            </ul>
            <ul class="list-group" style="margin-bottom:1px">
                <li class="list-group-item">
                    <a class="row" href="house_detail.jsp?id=19">
                        <span class="pull-left targe">[房间介绍] 测试房间</span>
                        <span class="pull-right date">2018-06-06   </span>
                    </a>
                </li>
            </ul>
        </div>
    </div>
    <div class="btn-group col-xs-offset-5">
        <ul class="pagination">
            <li><a href="house.jsp?pageCurrent=1">«</a></li>
            <li><a href="house.jsp?pageCurrent=1">1</a></li>
            <li><a href="house.jsp?pageCurrent=1">»</a></li>
        </ul>
    </div>
</div>
</div>
```

任务五　主要功能模块的实现

任务要求

本任务选择“会员管理模块”和“孕婴网主页”两个典型模块予以实现，根据软件开发的实际步骤，一步一步带领读者实现这两个典型模块的代码。

任务实现

（一）会员管理模块的实现

在本“孕婴网”中，会员管理主要包括：会员的登录、注册与基本信息修改。进行会员的注册与信息管理主要是为了提高会员服务质量，在以后的网站内容扩展上可以考虑“会员等级”“会员积分”

“会员奖励”“会员折扣”“会员购物”等相应的针对会员的服务功能。会员管理模块的登录功能的实现步骤为：

创建会员实体类 MemberInfo。在 src 下创建 com.cn.entity 包，在包下创建 MemberInfo 类，主要代码如下：

```
package com.cn.entity;
import java.util.Arrays;
import java.util.Date;
public class MemberInfo {
    private String id;
    private String memUsername;
    private String memPassword;
    private MemberLevel memberLevel;
    private String memName;
    private String memAge;
    private String memSex;
    private String memAddress;
    private String memTel;
    private String memPhone;
    private String memEmail;
    private Date regTime;
    private String cardNo;
    private String status;
    private Float memScore;
    private byte[] memPic;
    public MemberInfo() {   }
    public MemberInfo(MemberLevel memberLevel, String memName, String memPhone,
                    Date regTime) {
        this.memberLevel = memberLevel;
        this.memName = memName;
        this.memPhone = memPhone;
        this.regTime = regTime;
    }
    public MemberInfo(MemberLevel memberLevel, String memName, String memAge,
                    String memSex, String memAddress, String memTel, String memPhone,
                    String memEmail, Date regTime, String cardNo, String status,
                    Float memScore, byte[] memPic) {
        this.memberLevel = memberLevel;
        this.memName = memName;
        this.memAge = memAge;
        this.memSex = memSex;
        this.memAddress = memAddress;
        this.memTel = memTel;
        this.memPhone = memPhone;
        this.memEmail = memEmail;
        this.regTime = regTime;
```

```
        this.cardNo = cardNo;
        this.status = status;
        this.memScore = memScore;
        this.memPic = memPic;
    }
    public MemberInfo(MemberLevel memberLevel, String memName, String memAge,
                      String memSex, String memAddress, String memTel, String memPhone,
                      String memEmail, Date regTime, String cardNo, String status,
                      Float memScore) {
        this.memberLevel = memberLevel;
        this.memName = memName;
        this.memAge = memAge;
        this.memSex = memSex;
        this.memAddress = memAddress;
        this.memTel = memTel;
        this.memPhone = memPhone;
        this.memEmail = memEmail;
        this.regTime = regTime;
        this.cardNo = cardNo;
        this.status = status;
        this.memScore = memScore;
    }
    public String getMemUsername() {
        return memUsername;
    }
    public void setMemUsername(String memUsername) {
        this.memUsername = memUsername;
    }
    public String getMemPassword() {
        return memPassword;
    }
    public void setMemPassword(String memPassword) {
        this.memPassword = memPassword;
    }
    public String getId() {
        return id;
    }
    public void setId(String id) {
        this.id = id;
    }
    public MemberLevel getMemberLevel() {
        return memberLevel;
    }
    public void setMemberLevel(MemberLevel memberLevel) {
        this.memberLevel = memberLevel;
    }
```

```
    public String getMemName() {
        return memName;
    }
    public void setMemName(String memName) {
        this.memName = memName;
    }
    public String getMemAge() {
        return memAge;
    }
    public void setMemAge(String memAge) {
        this.memAge = memAge;
    }
    public String getMemSex() {
        return memSex;
    }
    public void setMemSex(String memSex) {
        this.memSex = memSex;
    }
    public String getMemAddress() {
        return memAddress;
    }
    public void setMemAddress(String memAddress) {
        this.memAddress = memAddress;
    }
    public String getMemTel() {
        return memTel;
    }
    public void setMemTel(String memTel) {
        this.memTel = memTel;
    }
    public String getMemPhone() {
        return memPhone;
    }
    public void setMemPhone(String memPhone) {
        this.memPhone = memPhone;
    }
    public String getMemEmail() {
        return memEmail;
    }
    public void setMemEmail(String memEmail) {
        this.memEmail = memEmail;
    }
    public Date getRegTime() {
        return regTime;
    }
    public void setRegTime(Date regTime) {
```

```
        this.regTime = regTime;
    }
    public String getCardNo() {
        return cardNo;
    }
    public void setCardNo(String cardNo) {
        this.cardNo = cardNo;
    }
    public String getStatus() {
        return status;
    }
    public void setStatus(String status) {
        this.status = status;
    }
    public Float getMemScore() {
        return memScore;
    }
    public void setMemScore(Float memScore) {
        this.memScore = memScore;
    }
    public byte[] getMemPic() {
        return memPic;
    }
    public void setMemPic(byte[] memPic) {
        this.memPic = memPic;
    }
    @Override
    public String toString() {
        return "MemberInfo{" +
                "id=" + id +
                ", memUsername='" + memUsername + '\'' +
                ", memPassword='" + memPassword + '\'' +
                ", memberLevel=" + memberLevel +
                ", memName='" + memName + '\'' +
                ", memAge='" + memAge + '\'' +
                ", memSex='" + memSex + '\'' +
                ", memAddress='" + memAddress + '\'' +
                ", memTel='" + memTel + '\'' +
                ", memPhone='" + memPhone + '\'' +
                ", memEmail='" + memEmail + '\'' +
                ", regTime=" + regTime +
                ", cardNo='" + cardNo + '\'' +
                ", status='" + status + '\'' +
                ", memScore=" + memScore +
                ", memPic=" + Arrays.toString(memPic) +
                '}';
```

```
    }
```

创建数据库操作类 MemberDao，实现数据库的增删改查功能。在 src 下创建 com.cn.Dao 包，在此包下创建 MemberDao 类，主要代码如下：

```
package com.cn.Dao;
import com.cn.base.BaseDao;
import com.cn.base.Util;
import com.cn.entity.MemberInfo;
import java.sql.Connection;
import java.sql.ResultSet;
import java.sql.SQLException;
import java.sql.Statement;

public class MemberDao {
    //向数据库中插入会员信息
    public void Insert(MemberInfo memberInfo) throws SQLException {
        Connection connection = BaseDao.getConn();

        try{
            Statement state = connection.createStatement();//容器
            String sql = "insert into member_info (MEM_USERNAME, MEM_PASSWORD, LEVEL_ID,
            MEM_NAME, MEM_AGE, MEM_SEX, MEM_ADDRESS, MEM_TEL, MEM_PHONE,
            MEM_EMAIL, REG_TIME, CARD_NO, STATUS, MEM_SCORE,MEM_PIC) value ('" +
                    memberInfo.getMemUsername() + "','"+
                    memberInfo.getMemPassword() + "','" +
                    memberInfo.getMemberLevel().getId() + "','" +
                    memberInfo.getMemName() + "','" +
                    memberInfo.getMemAge() + "','" +
                    memberInfo.getMemSex() + "','" +
                    memberInfo.getMemAddress() + "','" +
                    memberInfo.getMemTel() + "','" +
                    memberInfo.getMemPhone() + "','" +
                    memberInfo.getMemEmail() + "','" +
                    Util.changeDate(memberInfo.getRegTime()) + "','" +
                    memberInfo.getCardNo() + "','" +
                    memberInfo.getStatus() + "','" +
                    memberInfo.getMemScore() + "','" +
                    memberInfo.getMemPic() + "')";
            state.executeUpdate(sql);
        }catch (Exception e){
            e.printStackTrace();
            throw e;
        }finally {
            connection.close();
        }
    }
    //修改会员信息
```

```
public void update(MemberInfo memberInfo) throws SQLException {
    Connection connection = BaseDao.getConn();
    try{
        Statement state = connection.createStatement();//容器
        String sql = "update member_info set MEM_NAME='" +
                memberInfo.getMemName() + "',MEM_AGE='" +
                memberInfo.getMemAge() + "',MEM_SEX='" +
                memberInfo.getMemSex() + "',MEM_ADDRESS='" +
                memberInfo.getMemAddress() + "',MEM_EMAIL='" +
                memberInfo.getMemEmail() + "',MEM_PHONE='" +
                memberInfo.getMemPhone() + "',MEM_PASSWORD='" +
                memberInfo.getMemPassword() + "' where ID='" +
                memberInfo.getId() + "'";
        state.executeUpdate(sql);
    }catch (Exception e){
        e.printStackTrace();
        throw e;
    }finally {
        connection.close();
    }
}
//通过用户名在数据库中查询会员信息
public MemberInfo findByName(MemberInfo memberInfo) throws SQLException {
    Connection connection = BaseDao.getConn();
    try {
        Statement state=connection.createStatement();//容器
        String sql="select * from member_info where MEM_USERNAME = '" +
                memberInfo.getMemUsername() + "'";                //sql语句
        ResultSet rs = state.executeQuery(sql);        //将sql语句传至数据库，返回的值为一个字符
//集，用一个变量接收
        memberInfo = getList(rs);
        connection.close();
        return memberInfo;
    }catch (Exception e){
        throw e;
    }
}
//通过主键在数据库中查询会员信息
public MemberInfo findById(MemberInfo memberInfo) throws SQLException {
    Connection connection = BaseDao.getConn();
    try {
        Statement state=connection.createStatement();//容器
        String sql="select * from member_info where ID = '" +
                memberInfo.getId() + "'";                //sql语句
        ResultSet rs = state.executeQuery(sql);        //将sql语句传至数据库，返回的值为一个字符
//集，用一个变量接收
```

```
            memberInfo = getList(rs);
            connection.close();
            return memberInfo;
        }catch (Exception e){
            throw e;
        }
    }
    //取得全部会员列表
    public MemberInfo getList(ResultSet rs) throws SQLException {
        MemberInfo memberInfo = new MemberInfo();
        while(rs.next()){
          memberInfo.setId(rs.getString(1));
          memberInfo.setMemUsername(rs.getString(2));
          memberInfo.setMemPassword(rs.getString(3));
          memberInfo.setMemName(rs.getString(5));
          memberInfo.setMemAge(rs.getString(6));
          memberInfo.setMemSex(rs.getString(7));
          memberInfo.setMemAddress(rs.getString(8));
          memberInfo.setMemPhone(rs.getString(10));
          memberInfo.setMemEmail(rs.getString(11));
        }
        return memberInfo;
    }
}
```

创建 Servlet 类 LoginServlet，实现网页内容收集与业务流程。在 src 下创建 com.cn.servlet.login 包，在此包下单击右键，选择 Servlet 类型，名称为 LoginServlet，如图 8-21 和图 8-22 所示。

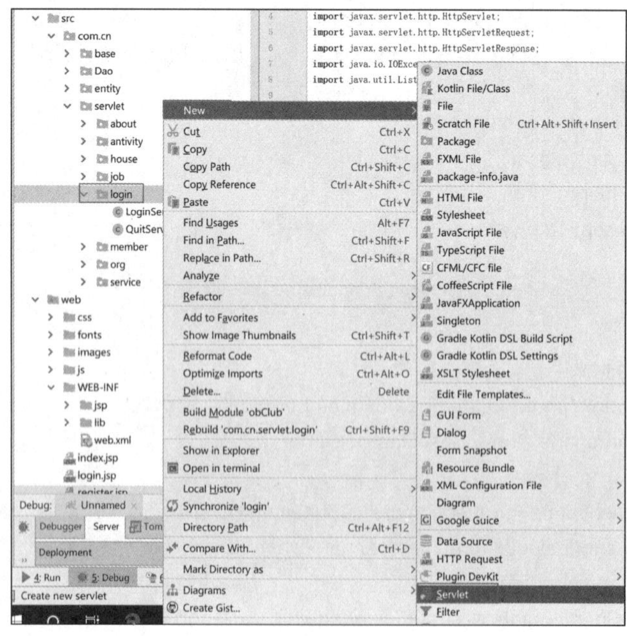

图 8-21

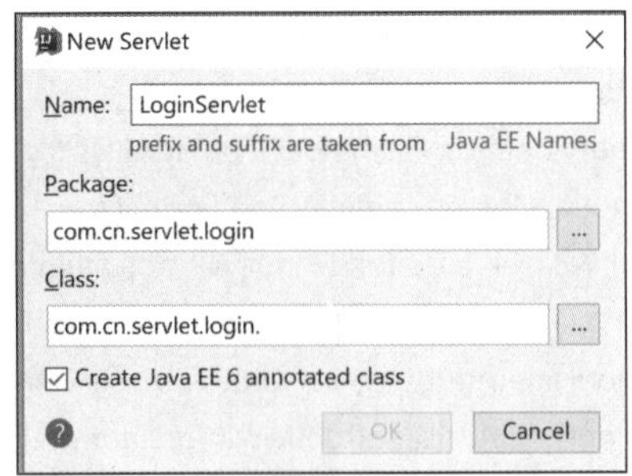

图 8-22

主要实现代码如下所示，在 Sevrlet 中主要实现 doGet 或者 doPost 方法，doGet 方法会自动响应页面使用 get 方式提交的请求，goPost 方法会自动响应页面使用 post 方式提交的请求，可以在响应请求的方法中实现两个目标，一是进行数据收集，二是进行业务流程的处理。在进行业务流程处理时可以调用实体类进行数据封装，可以调用数据库处理类进行数据处理。

```
package com.cn.servlet.login;

import javax.servlet.ServletException;
import javax.servlet.http.HttpServlet;
import javax.servlet.http.HttpServletRequest;
import javax.servlet.http.HttpServletResponse;
import java.io.IOException;
import java.util.List;
import com.cn.Dao.ActivityDao;
import com.cn.Dao.HouseDao;
import com.cn.Dao.MemberDao;
import com.cn.Dao.ServiceDao;
import com.cn.entity.ActivityInfo;
import com.cn.entity.HouseInfo;
import com.cn.entity.MemberInfo;
import com.cn.entity.ServiceInfo;

public class LoginServlet extends HttpServlet {

    @Override
    protected void doGet(HttpServletRequest request, HttpServletResponse response) throws Servlet
    Exception, IOException { }
    @Override
    protected void doPost(HttpServletRequest request, HttpServletResponse response) throws Servlet
    Exception, IOException {
        String message = "";
        request.setCharacterEncoding("GBK");
        response.setCharacterEncoding("GBK");
        try{
            HouseDao housedao = new HouseDao();
            ServiceDao serviceDao = new ServiceDao();
```

```
            ActivityDao activityDao = new ActivityDao();
            List<HouseInfo> houseInfos = housedao.findAll();
            List<ServiceInfo> serviceInfos = serviceDao.findAll();
            List<ActivityInfo> activityInfos = activityDao.findAll();
            if(houseInfos != null && serviceInfos != null && activityInfos != null){
              request.getSession().setAttribute("houseInfos",houseInfos );
              request.getSession().setAttribute("serviceInfos", serviceInfos);
              request.getSession().setAttribute("activityInfos",activityInfos);
            }
            //取得页面提交的用户名与密码
            String username = request.getParameter("username");
            String password = request.getParameter("password");
            //验证用户名与密码是否为空
            if (username == null || password == null){
              throw new NullPointerException();
            }
            //把用户封装到MemberInfo的实体类中
            //调用用户数据处理类MemberDao，通过findByName方法查询用户是否存在
            //根据查询结果进行页面跳转
            MemberInfo memberInfo = new MemberInfo();
            memberInfo.setMemUsername(username);
            MemberDao dao = new MemberDao();
            memberInfo = dao.findByName(memberInfo);
            if(memberInfo.getId() != null){
              if(memberInfo.getMemPassword().equals(password)){
                request.getSession().setAttribute("memberInfo", memberInfo);
                request.getRequestDispatcher("index.jsp").forward(request, response);
              }else {
                message = "登录失败";
                request.setAttribute("message", message);
                request.getRequestDispatcher("login.jsp").forward(request, response);
              }
            }else {
              message = "用户不存在";
              request.setAttribute("message", message);
              request.getRequestDispatcher("login.jsp").forward(request, response);
            }
        }catch (Exception e){
            e.printStackTrace();
            message = "登录失败";
            request.setAttribute("message", message);
            request.getRequestDispatcher("login.jsp").forward(request, response);
        }
    }
}
```

设置 web.xml。因为创建了 servlet 访问类，想从页面访问到 LoginServlet 类，需要在 web.xml 中对 servlet 映射进行配置，打开 web.xml，在文件中加入如下代码：

```
<servlet>
    <servlet-name>loginMain</servlet-name>
    <servlet-class>com.cn.servlet.login.LoginServlet</servlet-class>
</servlet>
<servlet-mapping>
    <servlet-name>loginMain</servlet-name>
    <url-pattern>/login</url-pattern>
</servlet-mapping>
```

此处 servlet-name 是在页面访问 servlet 的访问名称，servlet-class 是此 servlet 的全路径类名，为了访问简洁方便，需要配置一个 servlet 映射，这样一个名称的 servlet 就可以用在多个 url 页面路径中。所以配置<servlet-mapping>，并且把 url 的访问路径<url-pattern>配置为/login，在页面中只要访问/login，页面域就会自动导航提交到 com.cn.servlet.login.LoginServlet 进行处理。

创建用户登录页面 login.jsp。在 web->WEB-INF 下右键单击，选择 New->JSP/JSPX，创建 login.jsp 页面，如图 8-23 和图 8-24 所示。此页面的主要功能包括用户名与密码输入区域、新用户注册链接、登录提交按钮。页面展示如图 8-25 所示。

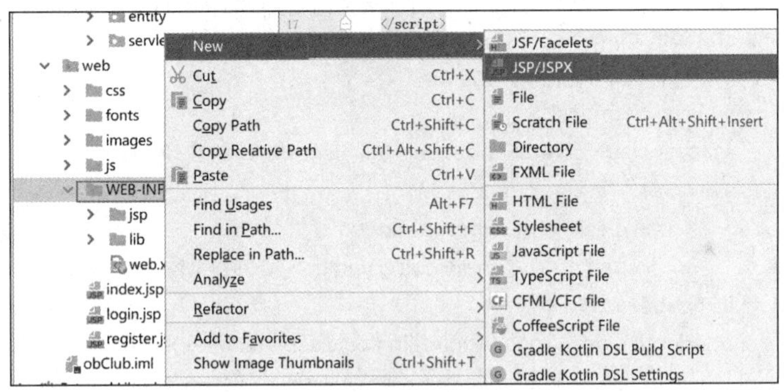

图 8-23

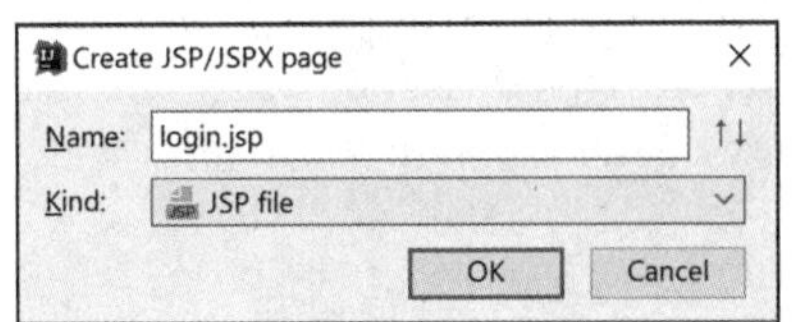

图 8-24

图 8-25

页面主要代码的实现如下所示：

```
<%@ page contentType="text/html;charset=GBK" language="java" %>
```

```
<html>
<head lang="en">
    <meta charset="GBK">
    <title>Title</title>
    <link rel="stylesheet" href="css/bootstrap.min.css"/>
    <link rel="stylesheet" href="css/web.css"/>
</head>
<body>
<jsp:include page="WEB-INF/jsp/index/index-header.jsp"/>
<%
    Object message = request.getAttribute("message");
    if (message != null && message!=""){
%>
    <script type="text/javascript">
        alert("<%=message%>")
    </script>
<%}%>
<div class="login" >
        <div class="col-lg-offset-4 col-md-offset-4 col-sm-offset-4 col-lg-4 col-md-4 col-sm-4"
        style="height: 300px;   border-radius: 5px; background-color: #EFEFEF; padding-top: 50px;
        padding- bottom: 15px">
            <form action="login" method="post">
                <div class="form-list">
                    <label class="control-label" for="txt_userId">用户名</label>
                    <div class="input-group">
                        <span class="input-group-addon">
                            <span class="glyphicon glyphicon-user"></span>
                        </span>
                        <input class="form-control" name="username" type="text" size="16" value=""
                        style="background-color: white"   placeholder="User ID" required/>
                    </div>
                </div>
                <div class="form-group">
                    label class="control-label" for="txt_password">密码</label>
                    div class="input-group">
                        <span class="input-group-addon">
                            <span class="glyphicon glyphicon-lock"></span>
                        </span>
                        <input class="form-control" name="password" type="password" size="16" value=""
                        style="background-color: white"   placeholder="Password" required/>
                    </div>
                </div>
                <div class="form-group" style="margin-top: 30px">
                    <input class="form-control btn btn-success" name="submit" type="submit" value="登
                    录"/>
                    <a href="register.jsp" style="float: right">现在注册>></a>
```

```
                </div>
            </form>
        </div>
</div>
<jsp:include page="WEB-INF/jsp/index/index-foot.jsp"/>
</body>
</html>
```

提示

本网站使用 Bootstrap 前端框架进行布局，读者在使用时把本项目提供的 css、font、images、js 文件夹及其文件夹下的资源文件，拷贝到 web 目录下即可拥有前端框架资源。在页面中使用<link rel="stylesheet" href="css/bootstrap.min.css"/>引入布局框架，读者也可以根据自己熟悉的技术进行页面布局。

由于顶部 Logo 导航和底部地址、电话信息，是整个网站中每个页面都有的内容，为了内容便于统一管理，所以单独制作了 Logo 导航顶部功能页面 index-header.jsp 和底部地址电话功能页面 index-foot.jsp，在本页 login.jsp 中，使用 include 动作标记引入页面：<jsp:include page="WEB-INF/jsp/index/index-header.jsp"/>和<jsp:include page="WEB-INF/jsp/index/index-foot.jsp"/>。

用户名、密码所在表单的 form 配置为<form action="login" method="post">，此处的 action 应与 web.xml 配置的路径的<url-pattern>/login</url-pattern>一致。

"会员管理"模块中的会员登录功能通过以上步骤已基本实现，读者可以根据实现步骤，自行练习"会员注册"功能以及"会员信息修改"功能，由于篇幅有限，此处不再赘述。

（二）"孕婴网"主页模块的实现

"孕婴网"主页是整个网站的灵魂，负责网站的企业形象、业务宣传、联系交流等。此网站主页的主要展示版块有：房间介绍、专业服务、套餐及活动。这三个版块，每个版块在主页的内容布局区域占一个单独部分，主要展示此版块的前四条图文信息。网站主页模块的主要功能实现步骤为。

创建"房间介绍""专业服务""套餐及活动"三个版块的数据访问 DAO 类。

在 com.cn.Dao 包下创建"房间介绍"数据访问 HouseDao 类，主要代码如下：

```
public class HouseDao {
    //取得所有房间信息列表
    public List<HouseInfo> findAll(){
        Connection connection = BaseDao.getConn();
        try{
            Statement state = connection.createStatement();//容器
            String sql = "select * from house_info";
            ResultSet resultSet = state.executeQuery(sql);
            return getList(resultSet);
        }catch (Exception e){
```

```
                e.printStackTrace();
                return null;
            }
        }
        //通过主键查找房间信息
        public List<HouseInfo> findById(String id){
            Connection connection = BaseDao.getConn();

            try{
                Statement state=connection.createStatement();//容器
                String sql = "select * from house_info where ID = " + id ;
                ResultSet resultSet = state.executeQuery(sql);
                return getList(resultSet);
            }catch (Exception e){
                e.printStackTrace();
                return null;
            }
        }
    //把房间信息封装到列表
        private List<HouseInfo> getList(ResultSet resultSet){
            try {
                List<HouseInfo> list = new ArrayList<>();
                while (resultSet.next()){
                    HouseInfo houseInfo = new HouseInfo();

                    houseInfo.setId(resultSet.getLong(1));
                    houseInfo.setHouseTitle(resultSet.getString(2));
                    houseInfo.setHousePrice(resultSet.getFloat(3));
                    houseInfo.setHouseDetail(resultSet.getString(4));
                    houseInfo.setHousePic(resultSet.getBytes(5));
                    houseInfo.setCreatedate(resultSet.getDate(6));
                    houseInfo.setShoworder(resultSet.getInt(7));
                    houseInfo.setIfshow(resultSet.getString(8));
                    list.add(houseInfo);
                }
                return list;
            } catch (Exception e) {
                e.printStackTrace();
                return null;
            }
        }
}
```

在 com.cn.Dao 包下创建“专业服务”数据访问 ServiceDao 类，主要代码如下：

```
public class ServiceDao {
    //取得所有专业服务列表
```

```
public List<ServiceInfo> findAll(){
    Connection connection = BaseDao.getConn();
    try{
        Statement state = connection.createStatement();//容器
        String sql = "select * from service_info";
        ResultSet resultSet = state.executeQuery(sql);
        return getList(resultSet);
    }catch (Exception e){
        e.printStackTrace();
        return null;
    }
}
//通过主键查找专业服务
public List<ServiceInfo> findById(String id){
    Connection connection = BaseDao.getConn();

    try{
        Statement state=connection.createStatement();//容器
        String sql = "select * from service_info where ID = " + id ;
        ResultSet resultSet = state.executeQuery(sql);
        return getList(resultSet);
    }catch (Exception e){
        e.printStackTrace();
        return null;
    }
}
 //把专业服务信息封装到列表
private List<ServiceInfo> getList(ResultSet resultSet){
    try {
        List<ServiceInfo> list = new ArrayList<>();
        while (resultSet.next()){
            ServiceInfo serviceInfo = new ServiceInfo();

            serviceInfo.setId(resultSet.getLong(1));
            serviceInfo.setServiceType(resultSet.getLong(2));
            serviceInfo.setSerTitle(resultSet.getString(3));
            serviceInfo.setSerDetail(resultSet.getString(4));
            serviceInfo.setSerPic(resultSet.getBytes(5));
            serviceInfo.setCreatedate(resultSet.getDate(6));
            serviceInfo.setShoworder(resultSet.getInt(7));
            serviceInfo.setIfshow(resultSet.getString(8));
            list.add(serviceInfo);
        }
        return list;
    } catch (Exception e) {
```

```
                e.printStackTrace();
                return null;
            }
        }
    }
```

在 com.cn.Dao 包下创建“套餐及活动”数据访问 ActivityDao 类，由于代码与 HouseDao 类和 ServiceDao 类基本雷同，读者可参考以上实现代码实现 ActivityDao 类，此处不再赘述，详细代码参考本项目完整示例。

创建主页 index.jsp 页面。在 web->WEB-INF 下创建 index.jsp 页面，页面主要版块功能区如图 8-26 所示。

图 8-26

每个版块显示四条图文信息，单击“更多”可进入此版块的二级页面。index.jsp 页面的代码如下：

```
<%@ page import="java.util.List" %>
<%@ page import="com.cn.entity.HouseInfo" %>
<%@ page import="com.cn.entity.ServiceInfo" %>
<%@ page import="com.cn.entity.ActivityInfo" %>
<%@ taglib prefix="c" uri="http://java.sun.com/jsp/jstl/core" %>
<%@ page contentType="text/html;charset=GBK" language="java" %>
<html>
<head>
    <title>Title</title>
</head>
<link rel="stylesheet" href="<%=request.getContextPath()%>css/bootstrap.min.css" />
<link rel="stylesheet" href="<%=request.getContextPath()%>css/web.css"/>

<body>
<%
    List<HouseInfo> houseInfos = (List<HouseInfo>)request.getSession().getAttribute("houseInfos");
    List<ServiceInfo> serviceInfos = (List<ServiceInfo>)request.getSession().getAttribute("serviceInfos");
    List<ActivityInfo> activityInfos = (List<ActivityInfo>)request.getSession().getAttribute("activityInfos");
    Object message = request.getAttribute("message");
```

```
        if (message != null && message!=""){
    %>
    <script type="text/javascript">
        alert("<%=message%>")
    </script>
    <%}%>
    <jsp:include page="WEB-INF/jsp/index/index-header.jsp"/>
    <div class="slider row">
        <div class="carousel slide" id="slider" data-ride="carousel" ng-controller="sliderCtrl">
            <ol class="carousel-indicators">
                <li class="active" data-target="#slider" data-slide-to="0"></li>
                <li data-target="#slider" data-slide-to="1"></li>
                <li data-target="#slider" data-slide-to="2"></li>
                <li data-target="#slider" data-slide-to="3"></li>
            </ol>
            <div class="carousel-inner" role="listbox">
                <div class="item active">
                    <img src="images/slider2.png" alt="" width="100%" height="400px" />
                </div>
                <div class="item">
                    <img src="images/slider4.png" alt="" width="100%" height="400px" />
                </div>
                <div class="item">
                    <img src="images/slider5.png" alt="" width="100%" height="400px" />
                </div>
                <div class="item">
                    <img src="images/slider3.png" alt="" width="100%" height="400px" />
                </div>
            </div>
            <a href=".slide" class="left carousel-control" role="button" data-slide="prev"> <span class=
            "glyphicon glyphicon-chevron-left" aria-hidden="true"></span> <span class="sr-only">
            Previous </span> </a>
            <a href=".slide" class="right carousel-control" role="button" data-slide="next"><span class=
            "glyphicon glyphicon-chevron-right" aria-hidden="true"></span> <span class="sr-only">next
            </span></a>
        </div>
    </div>
    <%if(houseInfos != null && activityInfos != null && serviceInfos != null){%>
    <div class="fast-nav">
        <div >
            <ul class="nav-ul">
                <li><a href="#">快速入口</a></li>
                <li><a href="service.jsp">专业服务</a></li>
                <li><a href="activity.jsp">套餐活动</a></li>
                <li><a href="house.jsp">房间介绍</a></li>
            </ul>
        </div>
```

```
</div>
<div class="container">
    <div class="more-list">
        <div class="more-list-header">
            <h4>房间介绍</h4> <small>room</small>
            <a href="house.jsp">更多>></a>
        </div>

        <div class="content" style="margin-top: 3px">
            <div class="row">
                <% for(int i = 0; i <= 3; i++){%>
                <div class="col-lg-3">
                    <div class="img-thumbnail" style="float: left">
                        <a href="house_detail.jsp?id=<%=houseInfos.get(i).getId()%>">
                            <img src="<%=request.getContextPath()%>/images/houseShowPic?id= <%=
                            houseInfos. get(i).getId()%>" style="height: 195px;width: 260px" class="img-
                            responsive" alt="" /></a>
                        <p><%=houseInfos.get(i).getHouseTitle()%></p>
                    </div>
                </div>
                <%}%>
            </div>
        </div>
    </div>
    <div class="more-list">
        <div class="more-list-header">
            <h4>专业服务</h4><small>server</small>
            <a href="service.jsp">更多>></a>
        </div>
        <div class="content" style="margin-top: 3px">
            <div class="row">
                <% for(int i = 0; i <= 3; i++){%>
                <div class="col-lg-3">
                    <div class="img-thumbnail">
                        <a href="service_detail.jsp?id=<%=serviceInfos.get(i).getId()%>&typeId=<%=
                        serviceInfos.get(i).getServiceType()%>"><img src="<%=request.getContextPath()
                        %>/images/serShowPic? id=<%=serviceInfos.get(i).getId()%>" style="height:195px;
                        width: 260px" class="img-responsive" alt="" /></a>
                        <p><%=serviceInfos.get(i).getSerTitle()%></p>
                    </div>
                </div>
                <%}%>
            </div>
        </div>
    </div>
    <div class="more-list">
        <div class="more-list-header">
```

```
            <h4>套餐及活动</h4> <small>active</small>
            <a href="activity.jsp">更多>></a>
        </div>
        <div class="content" style="margin-top: 3px">
            <div class="row">
                <% for(int i = 0; i <= 3; i++){ %>
                <div class="col-lg-3">
                    <div class="img-thumbnail" style="float: left">
                        <a href="activity_detail.jsp?id=<%=activityInfos.get(i).getId()%>"><img src="
                        <%=request.getContextPath()%>/images/actShowPic?id=<%=activityInfos.get(i).
                        getId()%>" style="height: 195px;width: 260px" class="img-responsive" alt="" /></a>
                        <p><%=activityInfos.get(i).getActTitle()%></p>
                    </div>
                </div>
                <%}%>
            </div>
        </div>
    </div>
</div>
<%}%>
<jsp:include page="WEB-INF/jsp/index/index-foot.jsp"/>
<script src="js/jquery.js"></script>
<script src="js/bootstrap.min.js"></script>
</body>
</html>
```

提示

这里要在页面显示三个版块的列表，所以在页面引入类<%@ page import="java.util.List" %>、<%@ page import="com.cn.entity.HouseInfo" %>、<%@ page import="com.cn.entity.ServiceInfo" %>和<%@ page import="com.cn.entity.ActivityInfo" %>。

使用直接调用相关 DAO 取得信息列表，例如：List<HouseInfo> houseInfos = (List<HouseInfo>)request.getSession().getAttribute("houseInfos");，此列表用于显示相关版块区域的图文信息。

通过页面的 JSP 循环语句，显示每个版块的图文详细信息。

<%for(int i = 0; i <= 3; i++){%><a href="house_detail.jsp?id=<%=houseInfos.get(i).getId()%>"><img src="<%=request.getContextPath()%>/images/houseShowPic?id=<%=houseInfos.get(i). getId()%>"style="height: 195px;width: 260px" class="img-responsive" alt=""/></a><p><%=houseInfos.get(i).getHouseTitle()%></p><%}%>

任务六　项目运行发布

任务要求

本任务主要实现项目整体的运行和配置实现，特别是在 IDEA 中运行环境的配置。

任务实现

（一）配置 IDEA 运行环境

配置 IDEA 下 Web 运行环境，首先在 IDE 环境的右上角单击下拉框，选择 Edit Configurations，如图 8-27 所示，出现项目配置对话框，如图 8-28 所示，首先单击左上角“+”图标，在 Add New Configuration 列表下找到 Tomcat Server，然后选择 Local（本地）配置。选择 Local 后，会出现配置界面，如图 8-29 所示，配置 Server 标签页，然后单击 Development 标签页，配置部署，如图 8-30 所示。

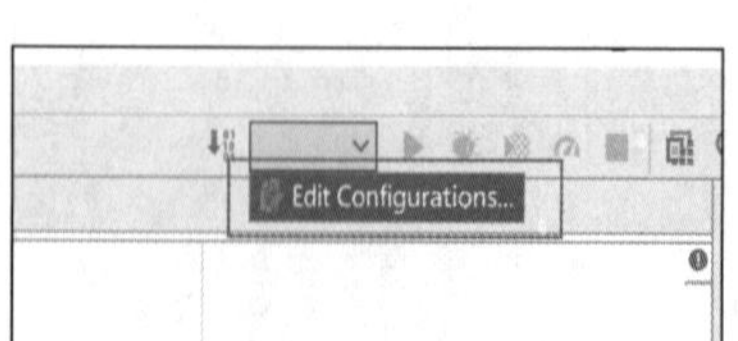

图 8-27

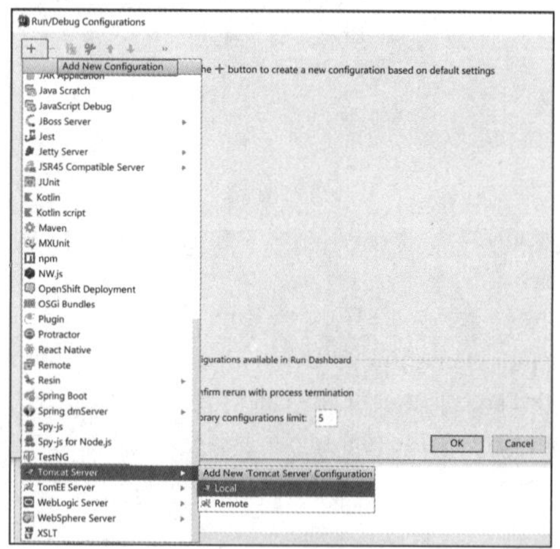

图 8-28

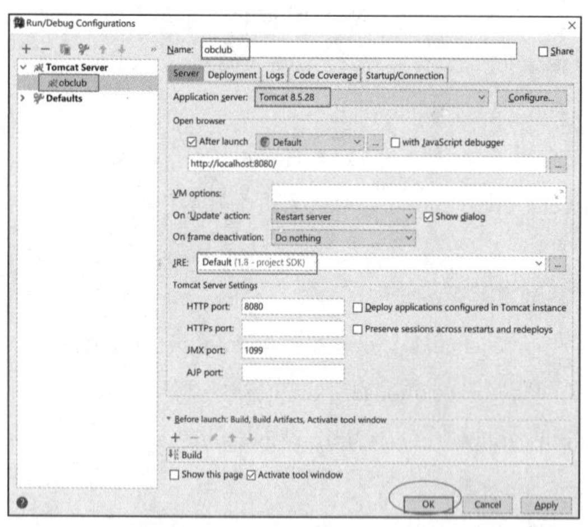

图 8-29

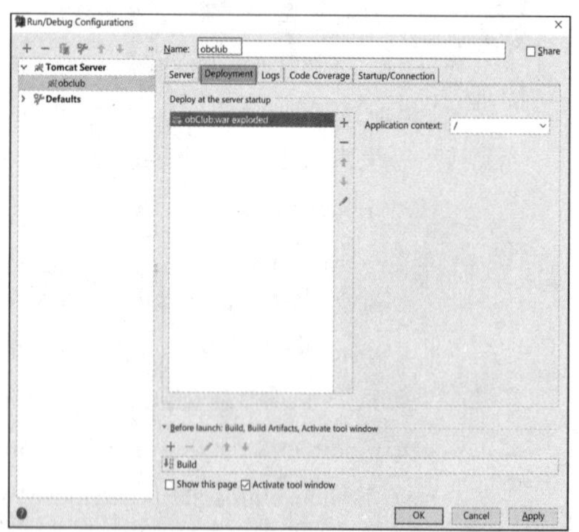

图 8-30

配置完成后，单击 OK 按钮即可。

（二）运行测试

运行 obclub 配置服务器，测试 login.jsp 页面。